FUTURE
METAL
STRATEGY

FUTURE METAL STRATEGY

Proceedings of an international conference
organized by The Metals Society and held
in the Europa Hotel, London,
on 10–11 May 1979

The Metals Society, London

Book 265
published in April 1980 by
The Metals Society
1 Carlton House Terrace
London SW1Y 5DB

Production editor: Derek Hatfull
Typefaces: IBM Century text with Avant Garde titles
Composition, printing and binding by
Inprint of Luton (Designers and Printers) Limited
Luton Bedfordshire

Contents

SESSION C: INFLUENCE OF DESIGN FACTORS ON
FUTURE UTILIZATION OF METALS

Professor J. Nutting, Conference Chairman
Opening remarks

May I welcome you to 'Future Metal Strategy' and give some slight background to this meeting.

Some two years ago as President of the Society I used the opportunity in my presidential Address to give a talk under the general title of 'Metals as Materials', and particularly to stress how society was very dependent on metals and therefore it did behove us to look very carefully at the use of metals, and how these might affect both the short- and long-term situations.

The theme that I had taken up was developed further in Mike Dowding's Presidential Address last year when he went on to discuss the world trade in metals; from this one began to see the important role that The Metals Society should take in looking at these problems of metal resources, trade in metals and metals use, at how they were all interlinked, and how they were going to develop.

As part of this interest, there is now a move to establish a materials or a metals forum. There are other learned societies that have similar interests and we hope that this forum, which will be jointly sponsored by the Institution of Mechanical Engineers, the Institution of Mining and Metallurgy, and The Metals Society, will begin to exert some influence on the particular issues of metal resources and metal use.

Having taken this initiative we then thought it would be appropriate to organize this meeting, in which we could start looking at a strategy; not, we might say, a very long-term strategy but a consideration of how the patterns of metals use would develop over the next 20 years. That is the theme of our meeting for these two days.

We are delighted to have with us, to give the introductory address, Sir Alan Cottrell. Appreciating the problems and demands on his time that he has to face in his duties as Vice Chancellor at Cambridge, we are especially pleased to welcome him because we remember with considerable interest his Hatfield Memorial Lecture, when he discussed some of the Malthusian problems that are besetting the world. We are well aware of his previous interests in these topics of metal use; and it is an interesting point (and one that I have tried to put over to people who are occasionally critical of the way that we teach metallurgy) that if one obtains a good grounding in physical metallurgy, this really stands one in tremendous stead for one's subsequent career. May I say I think there is no better example to justify this statement than Sir Alan, on whom I now call to give the opening address to this Conference.

1

Sir Alan Cottrell, FRS
Introductory address

It is a very great pleasure for me to be here this morning to say a few introductory words to this Conference on this extremely important subject of Future Metal Strategy.

I think that if we look at problems facing society today they appear quite bewildering and overwhelming; but if we stand back and try to get them into perspective they crystallize down, it seems to me, into two main problems, of which all the others are various consequences. The first is that of *natural resources*, the problem of providing future food, energy and materials in the world. The second is the *political* problem — the struggle between rival political schemes for the organization of society.

The problem of natural resources is open to systematic analysis, and I am going to talk about some aspects of that; but I ought to say first that it is hard to see how the general public and the politicians will really gather together the willpower to tackle the resources problem so long as they remain so much obsessed with the other problem, the political one. What is inescapable, I think, and a starting point, is that the world's population is continuing to rise (the most reliable extrapolations tell us that it is expected to double in the next 35 years). Considering that people, even today, are already in many parts of the world very inadequately fed, then it requires no

great imagination to see that there are immense problems lying in front of us.

But while the need for food grows roughly in proportion to the population of the world, need for energy and materials grows at a faster rate than that, for several reasons. Firstly, you need energy and equipment in order to produce food and the amounts you need go roughly in proportion to food — or rather, they rise somewhat faster than that because you have to work your land more intensively, which means more fertilizer and machinery on the land and things of that kind. Secondly, there are rising expectations of standard of living and there is economic growth, which together mean that the demands on an economy for materials and energy are growing faster than its population. Thirdly, there is the fact that as we dig away at leaner and leaner ores to get our materials and energy, then we have to consume more materials energy in doing that. So for all those reasons, the demand for materials and energy grows faster than food supplies; whereas the world's annual food demand has been doubling every 25 years or so, since the second world war the demand for metals has been doubling every 20 years and the demand for energy doubling every 15 years.

I admit that since the oil crisis of 1974 there has been a certain slipping back in the demand for energy and materials, but nevertheless it still remains intensive. Aluminium, for example, has been growing at a mighty 10% a year. As for steel, if we take the USA as a measuring point for the consumption of steel, the amount of steel in use in the USA at present is, I believe, equivalent to 10 tonnes per person there, and merely to keep that serviced by replacement, without any growth, would mean supplying one tonne per person per year just to maintain it. It seems very likely on present trends that most of the other OECD countries will move near to that same level of consumption in the next 20 years or so. But in other parts of the world, the position is very different. China has only about one-tenth of the USA consumption; in many parts of Africa, perhaps only about one-fiftieth. The picture, of course, is much the same for most of the metals — aluminium, copper and the rest — and we see that the essential world demand for all these is enormous over the next 20-30 years.

But what about supplies? We all remember the book that was produced a few years ago, 'Limits to Growth', which argued that, at present rates of consumption and allowing for the continuing growth, world resources of many of these things would soon be exhausted. It seems that metals such as mercury and tin will not be with us for many years more, except as rather expensive rarities, i.e. that mercury and tin will move into the gold and platinum class. Even the more common metals such as lead and zinc give cause for worry since, for quite basic geochemical reasons, it seems that there are no low-grade deposits of these metals — in other words, you go abruptly from a good ore to barren rock with nothing in between.

Incidentally, I wish the economists would understand these technical points, for they always make the bland assumption that as you lower the grade so the volume goes up as if by some law of nature. It is not a law of nature at all. Nature is not necessarily like that; it is a law of economists.

Having said that, I will admit that with copper and nickel there does operate something rather more like such a law, for there are large reserves of lower-grade ore. But the problem with nickel and copper will be partly environmental — one will have to work simply enormous volumes of rock in order to extract the mineral — and partly because of energy, for in working through these enormous amounts of rock one will consume colossal amounts of energy. Thus it is environment and energy which will be the problems for copper and nickel.

As regards aluminium, supplies of bauxite are still reasonably good but it does not look as if they are going to see us through for many years longer. Undoubtedly, sooner or later, we will have to turn to clay. There is of course no shortage whatsoever of that, but it will again mean yet more energy consumed in production.

I am not sure that I am allowed to mention iron but I will, just to point out that the position with iron is quite remarkable in that, going against the general trend, we are today actually able to use better grades of ore than we were using some years ago. The discovery of new first-class grades of ore has been so extensive that we are going in the opposite direction at the moment with iron.

Let me say a few words about energy, because we cannot get very far in these subjects without considering energy, since that is the fundamental resource. With energy, you can in principle solve all your other materials problems. If the worst came to the worst, you could extract all your minerals from seawater. I do not recommend it but in principle it could be done if you had enough energy. Without energy, none of the materials problems are solvable.

The demands on energy, as you well know, are increasing in all directions. Agriculture, as it becomes more intensive, is requiring more and more energy. A lot of modern Western agriculture is really a biological trick for turning fuel into food, and that process must go on as the world requirements for food increase. Purification of water is another example which is becoming increasingly necessary in some parts of the world, and that again is a great consumer of energy. Then the minerals themselves: as I mentioned a little while ago, the consumption of energy for the production of metals from ores in general is going up. I understand that, for example, in the USA the power used in metal mining rose fourfold in the 1950s just to keep the total mineral production constant. That is the sort of thing which is happening.

One of the obvious consequences of all this is that we must strive to make the best possible use of our materials. That not only means using them over again as far as possible, through conservation and recycling, but also means that we should develop the physical and chemical properties to the greatest possible extent, which imposes a continuing important task upon the physical metallurgist to develop these properties to their limits.

I would like to say a word in this respect about constructional materials, because these are the materials that are used in really large amounts. With the other things we can if necessary get by with more expensive sources, but for constructional materials we must require very large amounts of cheap constructional materials, and as energy becomes scarcer and more expensive it will eventually exert a dominant influence on the choice of constructional materials. I think we need to get ready for that change now. In fact, we can already see this happening in one or two cases. There are some metals such as magnesium,

titanium and vanadium which are really quite abundant, and by no means rarities. Yet although they are plentiful they are in fact only used in a rather specialist way because they require an enormous amount of energy in their production, and so their use as constructional materials has been largely inhibited by this great energy requirement. As we get to the end of the century, I think we shall begin to see that same cut-out starting to apply to some of the other materials.

If we look for a moment at energies of production for some of the common materials (the units I have, I regret to say, are the horrible SI units of gigajoules per tonne, and a 'giga' is 1000 million) for aluminium it is 330, which of course is mainly electricity; plastics 100; steel 45; and then cheap inorganic materials, by which I mean things like cement, concrete and glass, between 1 and 10. So there is an enormous spread in the energy requirements for the production of a tonne of various constructional materials. This does suggest strongly that in energy terms the future is likely to lie with the cheap inorganic materials. As of today their engineering properties have hardly been developed, mainly because they are just used in a very low-strength, bulk-space-filling sort of way. But that need not be so. Their intrinsic mechanical strength is extremely high. If you look at the elastic moduli, you will find that they are as high as any; and secondly, as we know from the fibre strengthening principle, you can build strong crack resistance into these materials. This is not ductility but crack resistance. The requirement that has always favoured ductility in metals has been crack resistance, but you can obtain crack resistance by an entirely different method through fibre reinforcement. So, scientifically, the way is open for the development of these cheap inorganics much more generally as strong construction materials. The attractiveness of that

5

is that these are materials that consume very little energy.

Lastly, a few words on policy for mineral resources. In the USA there is quite a serious effort to develop such a policy. To take just one example, a few years ago there was a very penetrating study by the National Academy of Science which was summarized in the report 'Materials and Man's Needs'. We do not seem to have anything comparable on this side of the Atlantic; indeed, if you want detailed data about world resources, metals and minerals, in general you have to go to the US for the prime sources.

I am reminded here of a remark made by an American Secretary of State who was criticizing some report that had been put on his desk. He said: 'It is like a British report, short on fact and long on argument'. I think this is true of our state of scholarship on the provision of materials throughout the world.

As regards policy, there was a start made a few years ago by Mr Heath at the beginning of his Government. He took quite an interest in materials supplies to this country, made an assessment of the vulnerability of the country to overseas supplies, and also spent some governmental money on assessing Britain's domestic metals resources. All this, of course, was in the early days of the North Sea oil discoveries; but apart from the North Sea oil and gas that venture withered as Mr Heath became embroiled in other problems, and really very little has happened since then.

There is today in the Department of the Environment quite a lively group which is studying limits to growth, but it is concentrating on future world trends and more general questions such as food supplies and so on. As regards specifically assessing the assurance of future materials supplies, this country is largely sailing blind today.

I think we could have hoped that the EEC might have taken up subjects such as this. Mineral supplies is surely an ideal subject for the EEC since all the Western European countries are really in the same boat in this respect. All of us in Western Europe are heavy importers of materials and we have a common interest in the security of our supplies; but regretfully it has to be said that the EEC does seem to be dominated by politicians and diplomats who tend to be myopic in regard to these problems.

Just to emphasize our vulnerability, let me remind you of the location of the main mineral reserves of some of our important metals. Chromium: 96% of the mineral reserves are to be found in South Africa and Rhodesia. Manganese: 83% in South Africa and the USSR. Vanadium: 94% in the USSR and South Africa. Tungsten: 72% in China, the USSR, and Korea.

It really is quite extraordinary that, after all the troubles we have had in recent years over OPEC oil and its supply, the countries of Western Europe should remain so complacent about these matters. That is why I am particularly glad to see this conference held at this time. I hope it will spread the story of our vulnerability as far as the politicians and perhaps get some political action taken.

SESSION A
General

P. C. F. Crowson
Geography and political economy of metal supplies

That the United Kingdom is heavily dependent on imports is a commonplace of any
discussion about future strategies for metals
supply. Whilst materials conservation and the
maximum amount of recycling consistent
with sound economics must figure prominently in any strategy, primary materials will in
most cases always be imported—and in
increasing quantities. This dependence both
constrains Britain's foreign policy options
and is also greatly influenced by the policies

---

*In this scene-setting paper the author ranges
over a very broad field and indicates some
potential problem areas without examining
them in depth. He then looks at the copper
industry in more detail, partly because
many generalizations are based on the copper
industry example, but also because the statistics both provide some support for the arguments about political instability as a barrier
to investment and also modify them. An
underlying theme of the whole paper is that
a more interventionist attitude to raw material procurement by HM Government in the
future than in the past is probably needed as
a response to external developments.*

*P.C.F. Crowson is with the Rio Tinto-Zinc
Corporation Ltd.*

adopted. Assurance of adequate supplies of
metal raw materials at competitive prices
should be a cardinal aim of government policy.
This paper starts from an examination of the
present position, and then examines some of
the problems likely to arise over the coming
decades. Wherever possible, it backs its argument with facts; too often it seems that much
of the conventional wisdom about raw material supplies is founded in myth.

THE PRESENT POSITION

In 1978, Britain's imports of metals totalled
£3 billion or rather less than 7½% of total
imports. Copper and aluminium each
accounted for around one-eighth of these
imports, and iron and steel for just over two-
fifths. The full make-up is shown in Table 1.

This table includes ores, concentrates,
waste, residues and scrap, primary unwrought
metals and semi-fabricated products. To the
extent that one metal may be imported in
finished form, whilst another is processed
within the United Kingdom, the relative
values of each metal must be treated cautiously. The value of metallic imports has trebled
during the decade, but has not risen as fast as
total imports, so that the share has declined
from over one-tenth. The average share of
metals in total imports has been 9·7% during
the 1970s. The value of imports fluctuates

9

Table 1   UK imports of metals in 1978

	£ million	%
Iron and steel	1 231	41·1
Copper	383	12·8
Nickel	101	3·4
Aluminium	380	12·7
Lead	109	3·6
Zinc	75	2·5
Tin	97	3·2
Silver and platinum group metals	278	9·3
Other non-ferrous ores and metals	200	6·7
Inorganic chemical elements, oxides and halogen salts*	140	4·7
TOTAL:	2 994	100·0

Source: Overseas Trade Statistics of the United Kingdom, December 1978; Divisions 28, 67, 68, 522 and 699·9 of the SITC (R2)
* Includes many metallic products, although part of the heading is more strictly chemicals

according to the level of economic activity within the UK, the amount of stockbuilding, and above all with metal prices. Thus the value of imports more than doubled between 1972 and 1974 during the commodity price boom. In 1978 the value was atypically low because of weak prices. Another yardstick is the share of metal imports in total final expenditure within the UK, which has averaged 1·7% during the 1970s and had dropped to 1·4% in 1978. Of course, this small share greatly underestimates the importance of metal imports to British industry.

The arithmetic of Britain's relative dependence on imports is worth emphasizing. Table 2 shows the share of imports in total consumption for eight major metals in the United Kingdom, the European Community as a whole, the United States, and Japan. The United Kingdom is slightly more dependent on imports of most of these metals (except lead and tin) than the European Community, but less dependent than Japan. All three, though, compare unfavourably with the United States. Even where the United States does rely heavily on imports, it can obtain a large share of its requirements by land from Canada or Mexico, tin and manganese excepted. But Britain, Japan, and the Community rely on much longer transport links,

Table 2   Import dependence in MAJOR metals (imports as percentage of consumption)

	UK	EEC	USA	Japan
Aluminium	62	61	85	100
Copper	82	81	15	90
Lead	46	53	13	76
Nickel	100	100	72	100
Tin	65	87	83	98
Zinc	100	68	59	80
Iron ore	89	79	36	99
Manganese	100	100	98	98

Sources: UK, EEC — Non-Fuel Minerals and Foreign Policy, RIIA, 1978; USA — Commodity Data Summaries 1979, US Bureau of Mines; Japan — Raw Materials and Foreign Policy, International Economic Studies Institute 1976

Notes: 1974–76 for UK and EEC, 1974 for Japan, 1976–78 for USA. Consumption includes secondary recovery. Figures allow for ores, concentrates and primary metals

Table 3   Import dependence in MINOR metals (imports as percentage of consumption)

	EEC	USA
Antimony	95	41
Cadmium	(90)	60
Chromium	100	91
Cobalt	100	98
Niobium	100	100
Germanium	100	33
Mercury	33	55
Molybdenum	100	—
Platinum group	100	91
Selenium	100	54
Tantalum	100	97
Titanium	100	n.a.
Tungsten	>99	52
Vanadium	99	33
Zirconium	100	n.a.

Sources: as Table 2. 1974–76 for EEC; 1976–78 for USA. In general all stages up to primary metal are covered by the table

usually by sea, with a wider spread of countries. Not only can these sea links be more easily disrupted, but many of their suppliers are potentially less stable than the United States' sources. The different positions of the United States and Europe are more clearly brought out in the following comparison of minor metals (Table 3).

The United Kingdom's dependence on imports of these materials is similar to that of the EEC, except that Britain has no indigenous sources of mercury. The differing degrees of dependence of Europe and Japan on the one

hand and the United States on the other raise the possibility, however remote, of the United States pursuing foreign policies which may conflict with their allies' interests. In this respect alone it is, perhaps, fortunate that the United States is almost as dependent as Europe on imports of certain products which are mainly produced in potentially unstable areas.

The major suppliers of the United Kingdom's metallic imports are listed in the next table (4).

The shares in the table are taken from trade

Table 4   United Kingdom import dependence

	Imports as % of consumption	Major suppliers (% share in brackets)
Bauxite	100	Ghana (72), EEC (5), other W. Europe (18)
Alumina	100	Jamaica (78), Guyana (12), EEC (3)
Aluminium	62	Canada (18), EEC (9), other W. Europe (57), E. Bloc (8), Ghana (7)
Copper	82	Zambia (29), Canada (20), Chile (20), EEC (10), Australia (4), other W. Europe (3)
Lead	46	Australia (70), Canada (20), Peru (3)
Nickel	100	Canada (87), Norway (4), Australia (3)
Tin	65	Bolivia (47), Nigeria (15), Malaysia (9), EEC (8), South Africa (5), Australia (4)
Zinc	100	EEC (31), Canada (20), other W. Europe (17), Peru (10), Australia (9)
Iron ore	89	Canada (20), Sweden (19), Brazil (15), Venezuela (9), Mauritania (9), Norway (7), USSR (6), Australia (4)
Manganese ore	100	South Africa (31), Brazil (29), Gabon (18), Ghana (8)
Antimony metal	100	EEC (92)
Cadmium	81	Canada (28), Japan (18), EEC (12), Australia (11), South Africa (9), E. Bloc (8)
Chromium	100	S.Africa (39), Philippines (27), EEC (4), USSR (3), other W. Europe (7)
Cobalt	100	No breakdown available, but mainly Zambia and Zaire
Niobium and tantalum	100	Brazil (82)
Mercury	100	EEC (34), Spain (20), USSR (25), Yugoslavia (4), China (3)
Platinum group	100	S. Africa (41), Japan (18), USA (18), USSR (6)
Titanium : ilmenite	100	Australia (68), Norway (31)
Titanium : rutile	100	Australia (99)
Tungsten	99·5	Portugal (21), Thailand (15), E.Bloc (14), Australia (7), EEC (6), Rwanda (5), Spain (5), Bolivia (5), Malaysia (5)
Vanadium (as V_2O_5)	100	S. Africa (55), Finland (42)
Zirconium	100	Australia (87), USA (7), EEC (2)
Gold	100	S. Africa (73), USA (9), EEC (2)
Silver	100	India (25), EEC (21), USA (20), Switzerland (12), Japan (7), South Africa (3)

Source: Non-Fuel Minerals and Foreign Policy Data Base, RIIA, October 1977

Note: Figures, based on 1974-76 averages, in most cases cover all stages of production up to primary metal, expressed in terms of metal content. Secondary recovery included in consumption only for main non-ferrous metals

11

Table 5 Major producers' percentage shares of world mine production and reserves (major metallic minerals)

	BAUXITE		COPPER		IRON ORE		LEAD		MANGANESE		NICKEL		TIN		ZINC	
	Prod.	Reserves	Prod.	Reserves	Prod.	Reserves	Prod.	Reserves	Prod.	Reserves	Prod.	Reserves	Prod.	Reserves	Prod.	Reserves
Australia	31·6	23·7	2·7	1·6	11·0	11·5	11·6	13·5	6·5	5·0	11·0	9·2	4·6	3·3	7·8	12·7
Canada	—	—	10·3	6·4	5·9	11·7	9·3	9·5			29·4	14·3			17·4	18·7
South Africa			2·6	1·1	1·0	1·2			23·1	33·3	2·8	—		0·4	0·5	1·3
United States	2·2	0·1	17·7	19·5	8·0	3·9	15·4	20·6			1·7	0·3			6·5	14·7
Japan			0·9	0·2	—		1·5	1·0							4·4	3·3
Western Europe	8·4	2·9	3·6	2·2	8·1	6·5	12·6	—			2·0	—	1·8	2·8	13·1	11·7
Total developed countries	42·2	26·8	38·0	31·0	34·0	34·8	50·4	57·1	29·6	38·3	46·9	24·8	6·4	6·5	49·7	62·4
USSR	5·6	0·7	11·0	7·3	27·8	30·1	17·4	12·7			17·4	9·5	14·1	6·2	16·5	8·0
China	8·7	3·4	8·6	4·7	7·5	2·9	4·0	}9·5					}8·6	15·0	}10·8	3·3
Eastern Europe					2·7	0·7	6·8									
Total Eastern Bloc	14·3	4·1	19·6	12·0	38·0	33·7	28·2	21·4	43·4	45·5	20·7	12·0	22·7	21·2	27·3	11·3
Bolivia													13·4	9·8		
Brazil	1·2	9·3			10·2	17·5			4·4	1·4			2·9	6·0	0·8	2·2
Chile			13·8	19·5												
Cuba											5·3	5·7				
Gabon									8·4	2·5						
Guinea	14·0	30·4														
Guyana	3·4	3·7														
India	1·8	5·9			4·9	6·0			8·0	1·0						
Indonesia	1·6										2·1	8·3	10·7	24·0		
Jamaica	13·7	7·4														
Liberia					2·1	0·7										
Malaysia													25·4	8·3		
Mexico							4·7	4·0							4·2	2·0
New Caledonia											12·7	25·0				
Nigeria													1·5	2·8		
Papua New Guinea			2·4	2·9												
Peru			4·5	6·4			5·2	3·2							7·6	4·7
Philippines			3·5	3·6							4·7	2·0				
Surinam	5·9	1·8														
Thailand													10·8	12·0		
Venezuela					1·7	1·4										
Zaire			2·8	4·7									1·4	2·0		
Zambia			4·1	6·7												
Total less developed countries	43·5	69·1	42·4	57·0	28·0	31·5	21·4	21·5	27·0	16·3	23·4	63·2	70·9	72·3	23·0	26·3

Notes: Sub-totals include shares of countries for which separate figures are not given, even where that country may be listed. Shares of production based on 1977-78 averages. Sources: *reserves* — US Bureau of Mines, Commodity Data Summaries 1978, 1979; Mineral Facts and Problems (Bicentennial Edition); *production* — as reserves and Metal Statistics 1967-77 (Metallgesellschaft)

statistics, which show the immediate rather than the original source of shipment. The European Community, Japan, and the United States often figure in the list as important suppliers but their production may, in turn, be based on imported raw materials. Thus the table to some extent underplays the UK's dependence on other sources. Although their importance has in some instances diminished with Britain's adoption of the European Community's Common External Tariff, Commonwealth countries are prominent suppliers. Britain's sources of imports often differ radically from those of other EEC members. For instance, Britain obtains nearly 30% of its copper supplies from Zambia, one-fifth from Canada, and one-fifth from Chile, with Zaire scarcely featuring as a supplier except indirectly as a source of feed for European refineries which supply 10% of the UK's needs. In contrast, about one-quarter of the other EEC members' supplies are from Zaire and only one-twentieth from Canada. Its commitments to the European Community could conceivably draw the United Kingdom into policies which might not always correspond with its own best interests, strictly conceived.

This argument cannot be pushed too far, as both the United Kingdom and other members of the European Community are part of a broader world market. The lack of tariff and other barriers to trade within the Community, in any case, largely integrates the United Kingdom's raw material supplies with those of other members. Developments in one part of the world market have repercussions throughout. Even if consumers are insulated from physical effects of a disruption to supplies through long-term contracts, the prices they pay will quickly reflect the disruption.

For most metals prices paid under long-term contracts are linked in some manner to world prices, whether they directly reflect prices on terminal markets such as the London Metal Exchange, or move less frequently in annual bargaining marathons.

The previous tables have presented a static snapshot of Britain's present sources of supply, and it may be argued that this is irrelevant to the next twenty years. Even though production and consumption are dynamic, with new sources and uses developing, old mines becoming exhausted, and some consumers finding alternative raw materials or products,

there is a considerable degree of inertia in patterns of production and trade. The times needed to establish new mines are long and lengthening, and new mines are often located in existing supplying countries. Thus, today's patterns of trade are a broad guide for the next decade at the very least. Nonetheless, the patterns of world production only imperfectly mirror the distribution of reserves, as Table 5 shows for eight major metals. The statistics are taken mainly from publications of the United States Bureau of Mines and should be treated as only approximate indications of the relative size of the major suppliers. The table shows only the more important producers of each metal and a blank does not necessarily indicate a lack of production or reserves. The sub-totals include the shares of countries for which no figures are given in the body of each section. Estimated reserves exclude deep-sea sources of manganese, nickel and copper.

Table 5 also brings out several interesting features of world mineral supplies such as the concentration of production and reserves in a relatively limited number of favoured countries; the tendency for Eastern Bloc countries to produce a greater share of the world total than their apparent reserves can sustain over the longer term (although the figures are especially suspect here); and the offsetting tendency for reserves of some metals in the less developed countries to be under-exploited. Yet in iron ore, lead, tin, and zinc, the shares of production, at least between broad geographical areas, are similar to the shares of reserves. The table clearly illustrates the widely diversified resource base of Australia, Canada, the United States, the USSR, and Brazil.

THE FUTURE

Ever since Britain's indigenous mineral supplies became inadequate to meet its total requirements in the late nineteenth century, the United Kingdom has relied on imports. There is also no evidence that its industrial output has been constrained by a lack of adequate supplies except when there has been a world shortage in which all consumers have suffered, others possibly more than the United Kingdom. On the contrary, it can be argued that too easy an access to raw materials in the UK has inhibited the development of substitutes which, in contrast, greatly assisted the development of Germany and Japan. Britain's slow pace of economic growth and, in most minerals, its declining absolute demands have enabled British industry to obtain ample supplies within the confines of long-established arrangements and with a minimum of government intervention. Governments of other more rapidly growing countries which lack such a diversified colonial past have needed to adopt a far more aggressive stance in order to ensure adequate metal supplies. Germany, France, and Japan, and even the United States, have seemingly shown much more concern about the adequacy of their raw material supplies than the United Kingdom, and have acted accordingly. This concern tends to wax and wane in intensity as materials become scarce or over-supplied. The Limits to Growth debate of the early 1970s, which coincided with the speculative excesses of the 1972-73 commodity price boom, and with the later oil price increases of 1973-74, reawakened fears that had lain dormant since the Korean War period. Fears about the immediate exhaustion of resources or the imminent cartelization of all mineral supplies have since been replaced by more sophisticated anxieties about a lack of investment to meet the potential demands of the later 1980s, or

about the terms on which supplies can be obtained.

We appear to be passing through a transition period from the relatively stable framework of the 1950s and '60s to a new post-colonial era. Whatever procurement arrangements may have served UK industry well in the past, there are grounds for developing new patterns, which would probably include a more active role for HM Government. To the extent that other important consumers do become more interventionist, British consumers may find themselves at a disadvantage in any competitive scramble for supplies. The speeding up of transport and communications links has eliminated the previously lengthy time lags between an event in some remote part of the world and its effects on supplies or demand. Instantaneous recognition of events has, perhaps, increased the importance of expectations and precautionary behaviour. More relevantly, countries appear to be more politically and economically interdependent than ever; events in one remote area soon impinge on the rest of the world. Another change is the decolonization of the Third World and the greatly diminished influence of the European metropolitan powers and the United States. The less developed countries soon realised that political independence did not bring economic freedom, and they have become far more assertive about economic control of their resources. Their resulting quest for a New International Economic Order is too deep-seated to be summarily dismissed, however unrealistic or extreme many of the component individual proposals might be. Acting largely through the Group of Seventy-Seven, within the United Nations and UNCTAD, the less developed countries have achieved an impressive degree of cohesion and bargaining power. They have been prepared to push the interests of all less developed countries even where these may conflict with the immediate needs of any one. Their passive reaction to the 1973-75 oil price increases illustrated this very clearly.

Table 6 Approximate shares of less developed countries in major metals (%)

	Reserves	Production	EEC imports	UK imports
Bauxite	69	44	34	75
Copper	57	42	67	55
Lead	22	21	30	5
Nickel	63	23	<10	< 5
Tin	72	71	92	85
Zinc	26	23	34	<15
Iron	32	28	50	<40
Manganese	16	27	40	30

Sources: Table 5 and Non-Fuel Minerals and Foreign Policy Data Base, RIIA 1978

Note: LDC's shares of reserves and production based on world totals including centrally planned economies. Latter excluded from LDC's.

As Table 5 has shown, the less developed countries' reserves of some major metallic minerals are relatively under-exploited. The data are summarized in the next table, which also indicates their importance as suppliers to the European Community as a whole, and to the United Kingdom, separately.

The supply of metallic commodities is only one aspect, albeit an important one, of the wider debate about international economic relationships. The governments of the industrial countries may be prepared to make concessions on commodities as part of much broader bargains, or in pursuit of political objectives. These concessions could easily pose problems for consumers of metals in the industrial countries. The United States' acceptance in principle of an International Copper Agreement in order to stave off a confrontation at UNCTAD V in Manila (and, perhaps, ease the burdens of some over-committed banks) is one instance. The evidence on the effectiveness of the International Tin Agreement is equivocal, although it has clearly been able to sustain market prices above a floor more easily than restrain them below a ceiling. Yet the tin industry is much smaller and more highly concentrated than copper. The buffer stock arrangement now being considered for the copper industry is unlikely to work satisfactorily; either it will prove to be an ineffective political gesture or other more restrictive arrangements will become necessary to preserve the value of the original buffer stock. The interests of all concerned might best be served by different arrangements from the outset.

Already many policy makers in the less developed countries are concentrating far more on the barriers, both real and supposed, to local processing. This will be a major subject of debate at UNCTAD V in Manila, and will figure prominently in international discussions in the next few years. There is already a growing tendency for the less developed countries to ship primary metal rather than ores and concentrates. In copper, for example, the less developed countries mined just over half the Western World's total in 1968 and 1977, but their share of refined metal output rose from some 23% to nearly 27% over the same period. Refined copper can be imported into the European Community duty-free, but there are tariffs on unwrought lead, zinc, and aluminium, and also on semi-fabricated products. Whilst the nominal tariff rates may seem modest, the effective rates on the value added at each stage can be significant — in the 20–30% range or even higher for many simple semi-fabri-

cated copper products, for example. The less developed countries can to some extent surmount these barriers by taking stakes in EEC fabricators as both Chile and Zambia have done in copper rod. This does not, however, achieve the LDC's objectives of increasing local value added and employment, and widening marketing opportunities through greater local processing.

There is a possibility that the less developed countries could make the acceptance of more highly processed products a condition of access to their raw materials, and there are already examples of this trend. Their ability to enforce such conditions of sale is increased by the spread of State ownership. In this regard Table 7 is illuminating. It compares the shares of State ownership and significant State Equity Interest in seven major metals with the less developed countries' shares of production in Western countries. A similar table for a year as recent as 1970 would have shown much smaller shares for public ownership, whilst in the early 1960s the bulk of Western metals production was privately owned. Not all State ownership is in the less developed countries, of course, and not all the mining industries of the less developed countries are State-owned. There is, though,

Table 7 Western world mine production: State ownership or significant Equity interest* and LDC share of production (%)

	Ownership, %	LDC share (%) of Western World production 1977-78
Bauxite	47·5 – 52·5	51
Copper	45·0 – 50·0	53
Iron ore	50·0 – 55·0	45
Lead	17·5 – 22·5	30
Zinc	22·5 – 27·5	32
Nickel	12·5 – 17·5	33
Tin	42·5 – 47·5	92

Sources: Table 5 and Stuart Harris's 'State Ownership in the World Mineral Industries,' CRES Working Paper, ANU Canberra

*Taken as Equity holding of 10% or greater
For copper, lead, zinc, and nickel the shares are of capacity, and for bauxite, iron ore, and tin of production. Copper excludes scrap, but tin includes some scrap. The figures are preliminary

a close correspondence between the two columns. There is no necessary relationship between the form of ownership and economic performance. Privately owned companies, for example, are seldom able to act without regard to wider considerations than shareholders' profits. Nonetheless, State-owned companies do often tend to pursue broader objectives than profitability. The narrow needs of a particular enterprise might be subordinated to considerations such as national employment, the balance of payments, or even diplomatic policies. To the extent that British industry is increasingly purchasing its metal supplies from State enterprises, it may need to call on the UK Government for assistance.

The cohesion displayed by the less developed countries in the UNCTAD discussions has been much less evident at the individual commodity level. The fears expressed in 1974-75 about commodity producers imitating OPEC have proved largely unfounded. With a few exceptions, there are too many producing countries, too wide a disparity between the costs of production in different areas, or too broad a range of substitutes for cartelization to succeed in raising prices significantly above the prevailing market level for extended periods. There is, nonetheless, a group of commodities which is produced either in a single country or is found in only a few places. Even if the likelihood of cartel-like

action is remote, a disruption of supplies could have the same adverse impact on availability and prices. Niobium is an example: most of the world's output comes from two or three mines, located in Brazil and Quebec. More frequently, fears are expressed about the Southern African dimension. Table 8 shows the extent of dependence on Southern Africa, defined as South Africa, Namibia, and Rhodesia. If Zaire and Zambia were included, copper and cobalt should be added to the list. Indeed, these Central African countries rely heavily on transport links to the south not only as an outlet for their products but even more to obtain fuel and supplies. Whilst Southern Africa is a relatively modest producer of major metals, it is the world's predominant exporter of platinum, manganese, and several steel alloying materials. The political pressures within and on Southern Africa mean that supplies from that source are becoming increasingly precarious. The problem is given added piquancy by two quirks of geology — the United States lacks indigenous sources of most of the minerals in question, and the USSR is the other major existing or potential producer. Table 9 compares the shares of world production in 1977-78 held by South Africa and the USSR.

It might be argued that these minerals are classic candidates for strategic stockpiling, if

Table 8   Dependence on Southern Africa: percentage shares of South Africa, Namibia, and Rhodesia

	Total world reserves	World mine production 1977-1978	EEC imports 1974-1976	UK imports 1974-1976
Copper	negligible	4	4½	2
Lead	neg.	1	4	negligible
Nickel	neg.	4	4	neglig.
Zinc	neg.	2	2	neglig.
Tin	neg.	2	1	5
Antimony	7	20	> 9	—
Arsenic	n.a.	21	n.a.	n.a.
Chromium	96	40	31	>31
Manganese	33	23	>52	31
Platinum group	73	46	40	40
Vanadium	19	39	30	30

Source: Non-Fuel Minerals and Foreign Policy Data Base, RIIA 1978; Commodity Data Summaries, 1979, US Bureau of Mines

Table 9   Shares of average 1977–78 world production

	South Africa	Eastern Bloc countries
Manganese	23	43 (USSR)
Chromium	34 (40 with Rhodesia)	32 (USSR and Albania)
Platinum group	46	46 (USSR)
Vanadium	39	31 (USSR)
Gold	58	21 (USSR)

Source: Commodity Data Summaries 1979, US Bureau of Mines

Britain wishes to release its foreign policy options from the constraints imposed by its dependence on Southern Africa. It is, though, impossible to amass strategic stockpiles quickly, and the decision should have been made several years ago. In any case, it is possible to draw widely differing conclusions about the implications of the Western world's reliance on South African supplies, depending on the basic assumptions made.

A more general issue for the future is the adequacy of investment today to meet expected demands. Much has been written elsewhere about these problems (e.g. 'Constraints to investment in minerals' by P.C.F. Crowson in *Transactions of the Institute of Mining and Metallurgy*, July 1978, and 'Mineral development in the eighties : prospects and problems': British North American Committee, November 1976). The coincidence of an exceptionally deep and prolonged recession in demand for most metals and consequent weak metal prices, on the one hand, with rapid capital cost inflation and the development of strong environmentalist pressures on the other greatly reduced both the incentive and the ability to invest in new mineral projects. These trends have been accentuated by the considerable change in favour of host countries in the fiscal and other terms under which such investment can take place. Private mining companies have been unwilling to accept the increased political risks of investing in less developed countries and unable to surmount the environmentalists' barriers to invest in the industrial countries. That at least is the conventional picture of recent trends; although reality appears both more prosaic and more complex. The one certainty is that broad generalizations are likely to be wrong.

THE EXAMPLE OF COPPER
The next three tables examine copper in some detail to assess the validity of these conclusions. The first of these, Table 10, compares the mine capacity in existence at the end of 1977 with prospective capacity in 1983. The 1977 base is built up from an analysis of all mines that produced copper in 1977 or were temporarily suspended at the end of the year. The 1983 total takes account both of subsequent closures and of new developments. Projects are included where firm construction plans have been announced and financing arranged. Both expansions to existing mines and completely new ventures are included. The table makes no allowance for potential new projects with short lead times that might be prompted by higher prices, nor for further cancellations or deferments. In 1978 approximately 15% of existing capacity was unutilized. The table brings out several interesting points, above all that mine capacity is still growing even after several years of supposed investment famine. It takes some time for large projects to be completed and these tide over a drop in new construction. Production would need to rise some 22% from the 1978 level to ensure that the mines in existence in 1983 run at an average 93% utilization rate. That requires an annual rate of growth of 4%, which is rather higher than many forecasters expect, particularly with a United States recession imminent. Nearly all the growth of capacity will be in less developed countries, including newcomers to the copper scene. Some, such as Iran, may be adjudged doubtful starters in 1979–80, but will almost certainly produce sometime in the next five years. The biggest uncertainty is not, however, over new projects but over the rate

Table 10   Present and prospective mine capacity for copper (end of period)

	1977		1983	
	10^3 tonne	%	10^3 tonne	%
USA	1645	23	1685	21
Canada	955	13	935	12
Australia	235	3	250	3
Japan	100	1	85	1
Europe	395	5	480	6
South Africa	225	3	230	3
Total OECD + S. Africa	3555	49	3665	46
Chile	1030	14	1090	14
Papua New Guinea	190	3	190	2
Peru	350	5	400	5
Philippines	325	4	440	5
Zaire	615	8	615	8
Zambia	760	10	760	9
Other less developed countries	425	6	855	11
Total less developed countries	3695	51	4350	54
Total Western world	7250	100	8015	100

Note: Detailed percentages may not add up to totals because of rounding. Source: RTZ Mine Information System

of utilization of existing capacity in Central Africa.

While political uncertainties do influence the direction of investment, the riskier areas also tend to be amongst the highest cost producers. This is illustrated in Table 11, which compares average head grades and production costs in the main copper-producing countries. The first column of the table shows the total production of each country in 1978, and the remaining three give information for 1977. Obviously, geological and geographical circumstances and the nature of mining in each country exert a powerful influence on costs. Grades are one of the major factors and the second column compares the average copper content of ores mined with the world average. Central Africa, Australia and Chile were well above average and Papua New Guinea, South Africa, and the Philippines well below.

To some extent these grades are offset by different mining costs as the third column indicates. These were highest in Australia and the Central African countries and lowest in Papua New Guinea and the United States. The costs of metal produced also take account of treatment and transport charges and of any by-products produced. The ranking in the final column is very different from that of the third. Large by-product credits in Canada, Australia and South Africa greatly improve those countries' relative positions, whilst the Philippines' position deteriorates. The highest cost areas are the United States and Central Africa. In the former, environmental costs are a powerful influence, whilst African political instability greatly boosts both operating and transport costs in Zaire and Zambia. Average metal costs in the countries not shown were collectively 13% above the world

Table 11   Characteristics of copper mining in major producing countries

	Production 1978, 10³ tonne	Head grade	Mining cost	Metal cost
		(percentage of total Western world)		
United States	1370	66	70	128
Chile	1035	170	95	85
Canada	680	73	127	59
Zambia	655	236	162	124
Zaire	420	490	152	103
Peru	345	120	96	98
Philippines	265	48	33	101
Australia	215	183	239	61
Papua New Guinea	200	64	53	79
South Africa	200	53	142	56
Western world (including countries not specified)	6120	100	100	100

Source: RTZ Mine Information System

Note: Head grades and costs are weighted averages, mining costs are operating costs per tonne of ore milled, and metal costs are cash breakeven costs per pound of contained copper metal produced net of all by-product credits

Table 12   Copper reserves of Western world: present mines and prospective projects (in terms of contained copper metal)

	Producing mines		Development projects		Exploration ventures	
	Reserves Mt	Grade, %	Reserves Mt	Grade, %	Reserves Mt	Grade, %
Australia	6·1	2·42	0·3	2·15	4·9	1·52
Canada	15·5	0·71	0·1	1·08	17·4	0·48
United States	52·4	0·69	2·1	1·12	25·0	0·74
South Africa	4·1	0·33	0·2	0·45	0·2	0·60
Japan	1·5	1·51	—	—	negl.	0·89
Western Europe	7·7	0·73	3·6	0·46	1·9	0·74
Total OECD and South Africa	87·3	0·7	6·2	0·6	49·4	0·65
Chile	73·1	1·07	2·3	0·74	16·9	0·69
Papua New Guinea	3·4	0·45	—	—	7·2	0·57
Peru	8·5	0·95	0·6	1·4	26·0	0·80
Philippines	9·5	0·46	2·7	0·53	2·8	0·46
Zaire	10·4	3·62	—	—	6·6	4·46
Zambia	26·0	2·95	—	—	8·4	7·52
Other less developed countries	20·0	0·86	18·0	1·31	56·2	1·54
Total less developed countries	151·0	1·08	23·5	0·88	124·2	0·78
Total Western World	238·3	0·9	29·7	0·81	173·6	0·74

Source: RTZ Mine Information System

Notes: The grades in this table are the average for all reserves and the percentages in Table 11 refer to the average grade of ore treated in 1977. Figures are taken mainly from company reports which seldom give total recoverable reserves

average. A rational procurement strategy for the United Kingdom would be to purchase from the lower-cost areas whilst seeking means of reducing any external political obstacles to efficiency in the higher-cost areas. An international copper agreement with a large buffer stock is more likely to ossify existing cost structures than to change them.

The final table (12) shows how exploration and development interest is concentrated on the higher-grade and lower-cost areas such as Chile, or in less developed countries without a large established copper industry. The first two columns show the contained copper metal reserves and their average copper contents for all producing mines. In total the less developed countries controlled 63% of the stated reserves of these mines and had much higher average grades. The average grades of projects now under development are lower than those of producing mines with some significant exceptions, most notably in North America. Less developed countries have 79% of the reserves of development projects, and 72% of those of exploration ventures. Whilst it would be wishful thinking to assume that all exploration ventures will be quickly turned into producing mines (many have been under study for many years), the table places statistics on exploration expenditures in perspective. There appears to be no shortage of potential projects, given the right economic circumstances for their development. Similar conclusions apply to the other major non-ferrous metals. In lead, for example, stated reserves of producing mines total 47·5 million tonnes of contained lead, there are 9·3 million tonnes under development, and a further 25 million tonnes in exploration ventures. The average grades of the latter are similar to those of producing mines.

W. O. Alexander
Total energy content of some metals and materials and their properties

Since we can assume that about half the world's energy utilized per annum is used in making metals and materials, it is proposed to outline some interrelating factors which will affect our future materials strategy. This view of tonnage materials for engineering and structural applications looks at metals and

Accurate assessment of the total energy contents of some tonnage metals and materials is proving difficult and variable. There is no direct relationship between selling prices and total energy for a given product. Two unresolved problems in conventional energy accounting are the recycling of new and old scrap and the justification for using single-life-cycle materials and composites. It is proposed that new parameters related to cost and total energy per unit of property must be used in the longer term to apportion the usage of energy for the manufacture of materials. Some future trends are suggested, based on the foreseeable relationship between energy and materials: availabilities, properties, life cycles, and recyclability.

The author is in the Department of Chemistry at The University of Aston in Birmingham.

plastics, and other non-metals such as concrete and timber, and evaluates likely developments in the light of energy and material availability and their properties.

ASSESSMENT OF TOTAL ENERGY CONTENT OF A MATERIAL

Publication of energy utilization data in metallurgical and material processing operations has been widespread in recent years. However, such data are in rather an early stage of compilation. In some cases there are discrepancies of 200–300% between values in different countries and between different firms. Recent work in the UK iron foundries[1] and aluminium industry[2] is also revealing that energy auditing in routine metallurgical establishments covering extended periods is up to double the values hitherto assumed. Difficulties have been experienced in assessing true works production data in total energy increments at each step. This is partly because metering of individual production units on a complex works site is not carried out, while other errors stem from various assumptions about overall energy usage on a site, e.g. space heating, furnace standby losses, etc. The overall estimate of the total energy which has already been incurred by an ore as raw material, concentrate, or virgin metal when it has arrived in the

United Kingdom is also not readily calculated. Since there are now very few ore deposits in the UK this is an important factor. Another factor which is not revealed in UK Department of Energy statistics is the quantity of bunker fuel shipped in foreign or British ports for bulk transportation of incoming raw materials.

Considerably more detailed work and agreed conventions will be necessary on an international basis before total energy data on materials are truly comparable. The Energy Audit Series of investigations by the Departments of Energy and Industry are probably the most accurate detailed reviews of energy usage in industry in the UK, but so far they have only covered iron castings, bricks, dairy products, and bulk refractories.*

ENERGY AND COSTS

It cannot be denied that market forces will in the end predetermine usage of energy and materials in whatever form, but this will only happen when energy costs become a major proportion of the cost of any end product. In some cases energy already accounts for 30% of the product's total value, but since energy expenditure on incoming raw materials earlier in the processes is not quantified in separate financial terms most managers and directors believe that the energy content in their part of the process only represents 10–15% of their total costs. Rigorous energy auditing would reveal a much higher value and, indeed, one author[3] would suggest that energy is one fundamental economic measure. One could argue that energy and labour make up almost the total cost of any metal or material.

In the interim, i.e. the next twenty years, it seems that total energy content in a product will not be accurately reflected in its cost. This is partly because the total energy content of a raw material is not known or appreciated; partly because various energy sources are so vastly different in basic costs[4] and hidden by state subsidies; and partly because energy

---

*Since the presentation of this paper at the conference, further reviews have been issued covering glass, aluminium, pottery, and coke; see Ref. 12 [*Author*]

auditing as yet cannot fairly apportion the energy used on a works site to each product group. There is also considerable discrepancy amongst observers in the field as to whether or not process energy varies significantly with output. Evidence within the occupacity 55 – 95% points in both directions in metallurgical plants.[5, 6] At a recent international energy/resources conference in the USA it was generally agreed that total selling prices of materials or products do not reflect their total energy contents.[7]

All these observations therefore justify the need to accurately assess and discuss the relevance of the total energy content of the major tonnage materials of the world.

UNRESOLVED CONVENTIONS FOR RECYCLING TOTAL ENERGY ASSESSMENT

The conventions for assessing total or gross energy seem to be fairly well accepted in Europe[8, 9] and this paper is not intended to define or elaborate upon them. There are, however, two other aspects of total energy auditing which will need general agreement and implementation. These are related to scrap recycling and the justification for manufacturing single-life-cycle and/or dissipated materials.

The first is the value of energy to be assessed as used in the recycling of new and old scrap. This obviously consists of two components:

(i) the inherent total energy originally used to make it the first time, and
(ii) the 5 – 10% additional energy required to recycle it to the place of remanufacture.

Most firms would like to assess it at (ii) only, thereby gaining (i) for nothing. On the other hand, all scrap metals have an intrinsic financial worth which is fairly close to their raw metal values, with the exception perhaps of iron and steel which is artificially low. This financial value of scrap metal contains the total energy content of its original extraction and manufacture. It would obviously then be anomalous to ignore its total energy content when it is recycled, since thereby lower total energy values would be obtained for the new processed product. This view is further supported by the argument that all the energy utilized in initial manufacture is a world energy asset and must not be destroyed. References 8 and 9 suggest that energy credits be given for recycled material.

The contrary view that on second and subsequent reprocessing much less additional or process energy is used per unit of product is attractive for short-term marketing but is not tenable by strict energy auditing standards.

The second problem is the vexed question of composites. These can be defined for this particular problem as any geometric arrangement of two or more materials which cannot easily be separated into their component materials without excessive expenditure of energy or with low yields. This wide definition would cover such products as steel-cored tyres, glass-reinforced plastics, bimetals, motorcar radiators, honeycomb structures, laminates, carbon filaments, 'Hifill', etc. In all such cases, the main reason for their manufacture is that in service they contribute significantly to savings in operating energy or savings made through longer service life.

Since by their very nature and manufacture they are unlikely to be recycled and are therefore lost, the energy used in making them must be justified. This can only be done by equating the energy used in their manufacture against the operating energy likely to be saved over the lifetime of the product. At the very lowest these two values should be equated, i.e. total energy used in complete manufacture to finished product = process operating or recurring energy saved over the average life of the product in service. If the operating energy saved is greater than this, then so much the better; if it is less, then the composite product must be justified for other valid reasons. Thus a composite or other single-life-cycle material might be justifiable if

(1) it gave the complete article a longer life
(2) the energy expended to make it was less than for a single material (but this is very unlikely)
(3) a unique technology was involved with no alternatives, e.g. microcircuitry and solid-state materials
(4) maintenance was significantly reduced.

It should be noted that this rigorous analysis will also be required for other single-life-cycle materials such as metals which are dissipated and plastics which are not recycled. Again, present IFIAS convention would give no energy credit for waste, and only enthalpy is included 'if the material is recycled'.[8,9] Consideration of the properties along with the total energy involved in manufacture give data which are absolute and likely to be immutable, i.e. when the total energy contents have been more refined and generally accepted we have data on which future trends can be estimated with greater confidence than prognostications based on current costs, monetary values, subsidies, etc. This thought is possibly of greater significance in the UK than in many other countries because our reserves are higher in energy and lower in metal and materials than those of most other countries. It is vital, therefore, to exploit this resource for maximum long-term benefit of the country and its inhabitants. Several countries already smelt aluminium with spare energy, some hydroelectric and some flare gas, and in this way export energy; but our energy in the UK is far from cheap and we would have either to make far more sophisticated products of low additional total energy or to develop new routes which are much more efficient in energy utilization.

TOTAL ENERGY
PER UNIT OF PROPERTY

Although we can evaluate the total energy consumed per kilogramme of finished material, such information does not convey the inherent value of the product to prospective users. The concept of appraising the value in total energy terms for the range of properties is of equal if not greater relevance. An outline of the properties of tensile strength, modulus of rigidity, and fatigue strength for some common metals and materials is given in Table 1, together with their specific energy, i.e. total energy per kilogramme of material; from these data the energy per unit of property can be readily calculated, as in columns 6 – 8. As would be anticipated, total energy criteria throw a completely different light on the true values of some materials to mankind.

Table 1 Energy consumption related to material properties

Material	Tensile strength, MN/m²	Modulus of rigidity, MN/m²	Fatigue strength, MN/m²	Density, kg/m³	Specific energy, kWh/kg	Total energy per unit of MN/m		
						Tensile strength	Mod. of rigidity	Fatigue strength
CAST IRON castings	400	45 000	105	7 300	4·0–16·0	73–292	0·65–2·60	278–1112
STEELS:								
EN1 low-alloy free cutting bar	360	77 000	193	7 850	16·0	349	1·63	651
EN24 1·5Ni–1Cr–0·25Mo bar	1000	77 000	495	7 830	16·0	125	1·63	253
Stainless 304 18Cu–8Ni sheet	510	86 000	250	7 900	32·0	229	2·94	487
NON-FERROUS METALS:								
Brass 60Cu–40Zn bar	400	37 300	140	8 360	27·0	565	6·05	1612
Aluminium alloy sheet	300	26 000	90	2 700	79·0–83·0	711–747	8·2–8·6	2370–2490
Duralumin sheet	500	26 000	180	2 700	79·0–83·0	427–449	8·2–8·6	1185–1245
Magnesium alloy bar	190	17 500	95	1 700	115·0	1029	11·17	2058
Titanium 6Al–4V bar	960	45 000	310	4 420	155·0	715	15·2	1520
PLASTICS:								
Propathene GWM22	35	1 500	7·5	906	20·0–40·0	517–1034	12–24	2400–4800
Polythene L.D. XRM21	13	84	3·25	920	15·0–30·0	1062–2124	165·0–330	4246–8492
Rigidex 2000	30	1 380	4	950	15·0–30·0	475–950	10·3–21	3563–7126
Nylon 66 A100	86	2 850	20	1 140	50	663	20	3400
PVC (R)	50	1 680	12·5	1 400	20·0–50·0	560–840	17·0–25·0	2240–3360
REINFORCED CONCRETE beam	38	10 000	23	2 400	2·3–4·0	145–253	0·55–0·96	240–417
TIMBER:								
Hardwood	14	4 500	6	720	0·5	26	0·08	60
Softwood	5	2 000	3	550	0·5	55	0·14	92

Table 2 Cost related to material properties (costs at 1 June 1979)

Material	Tensile strength, MN/m²	Modulus of rigidity, MN/m²	Fatigue strength, MN/m²	Density, kg/m³	Cost, £/tonne	Cost per unit of MN/m, £		
						Tensile strength	Mod. of rigidity	Fatigue strength
CAST IRON castings	400	45 000	150	7 300	500	9·1	0·08	24·3
STEELS:								
EN1 low-alloy free cutting bar	360	77 000	193	7 850	244	5·3	0·020	9·9
EN24 1·5Ni–1Cr–0·25Mo bar	1000	77 000	495	7 830	400	3·1	0·04	6·3
Stainless 304 18Cu–8Ni sheet	510	86 000	250	7 900	1 500	23·2	0·140	4·7
NON-FERROUS METALS:								
Brass 60Cu–40Zn bar	400	37 300	140	8 360	1 000	20·9	0·230	59·7
Aluminium alloy sheet	300	26 000	90	2 700	1 000	9·0	0·100	38
Duralumin sheet	500	26 000	180	2 700	2 000	11	0·210	30
Magnesium alloy bar	190	17 500	95	1 700	3 700	32	0·350	66
Titanium 6Al–4V bar	960	45 000	310	4 420	11 000	50	1·100	157
PLASTICS:								
Propathene GWM22	35	1 500	7·5	906	560	14·5	0·340	68
Polythene L.D. XRM21	13	84	3·25	920	535	38	—	151
Rigidex 2000	30	1 380	4	950	515	16	0·350	122
Nylon 66 A100	86	2 850	20	1 140	1 705	23	0·680	97
PVC (R)	50	1 680	12·5	1 400	445	12·5	0·370	50
REINFORCED CONCRETE beam	38	10 000	23	2 400	20	1·26	0·005	2·1
TIMBER:								
Hardwood	14	4 500	6	720	400	19	0·060	45
Softwood	5	2 000	3	550	250	27	0·070	46

For example, timber uses far less energy for a given strength, namely 26 – 55 kWh/MN, than any other material. Reinforced concrete is attractive at 145 – 250 kWh/MN, followed by steels at 125 – 350, while cast irons can vary from 300 to 1825 in the UK according to the energy efficiency of the manufacturer. The newer materials, aluminium, plastics, and titanium all use total energy contents of 400 for duralumin to 700 for other aluminium alloys, and 500 – 2000 for plastics, depending on the type of polymer and whether the energy content of the feedstock is included or not. Roughly the same order of merit obtains for modulus of rigidity and fatigue strength.

Table 2 looks at the same materials and considers their cost, which appears in column 5. By a similar set of calculations one can evaluate the cost per unit of property. For many of the materials this results in a somewhat similar ranking, i.e. the reinforced concrete, then the steels and cast irons, followed by a range of polymers, with aluminium alloys competing with plastics. However, there are a few drastic shifts in the ranking table — timber falls from cheapest in energy to almost dearest in cost terms. Two other marked shifts downward in ranking occur for stainless steel and titanium, possibly because of the high costs of getting good surface finishes on sheet. With the fall of these three materials to a high cost per unit of property, most of the other materials move up but maintain the relative ranking order.

This type of analysis can be extended to cover other properties of engineering and life performance. Furthermore, by a system of weighting and scaling the significance of a property in the overall performance sense, it is possible to determine the cheapest material in total energy or in cost terms for any predetermined combination of properties. Such estimates can be readily carried out by computer.[13]

FUTURE TRENDS

This brief review is more concerned with the total energy/metal (material) interface and the repercussions of this consideration. As a general indication of present thinking on the extraction side one can hardly do better than to bring out some observations made by

Professor H. H. Kellogg in his Julius Wernher Memorial Lecture.[10] This highlights the fact that true conservation saves energy and capital and the environment. He also illustrates the relative inefficiency of using electricity as compared with fossil fuel in many metallurgical processes, with the possible exception of electrolytic processes. This is also true of the capital cost of the different energy sources in dollars per GJ year. Similarly, hydrometallurgical processes do not look as attractive as pyrometallurgical routes, particularly where tonnage oxygen can accelerate the process, increase productivity and yield, and conserve capital, thereby saving 30–40% of energy. Yet another important aspect which he touches on under the heading of industrial symbiosis is the co-generation of steam and electric power, waste heat recovery, and environmental dust and fume control. All of these have been achieved in the past in various isolated cases across the world but from now on, with the increasing costs of power, they will merit more profound examination, particularly in the location of new industrial extraction processes.

AVAILABILITY
Energy

Another serious factor to bear in mind is that as the resource becomes more difficult to discover and win, so the total energy required per tonne of finished product rises. This is well exemplified in the case of lead. Figure 1[11] indicates that below a certain lead content in an ore body the energy cost becomes prohibitive, resulting in a decrease in lead supplies. One redeeming feature of this difficulty, which could counterbalance the problems caused by it, would be that it might encourage the recycling of lead already extracted. This can be done with the expenditure of one-tenth to one-twentieth of the energy required to produce it in the first place. But there is a snag to this also, as will be indicated later in this paper.

Although most estimates strictly in terms of 'reserves' show a levelling off in the world's energy reserves and supplies and the outlook is not good, nevertheless, as prices rise energy demands tend to fall and other resources

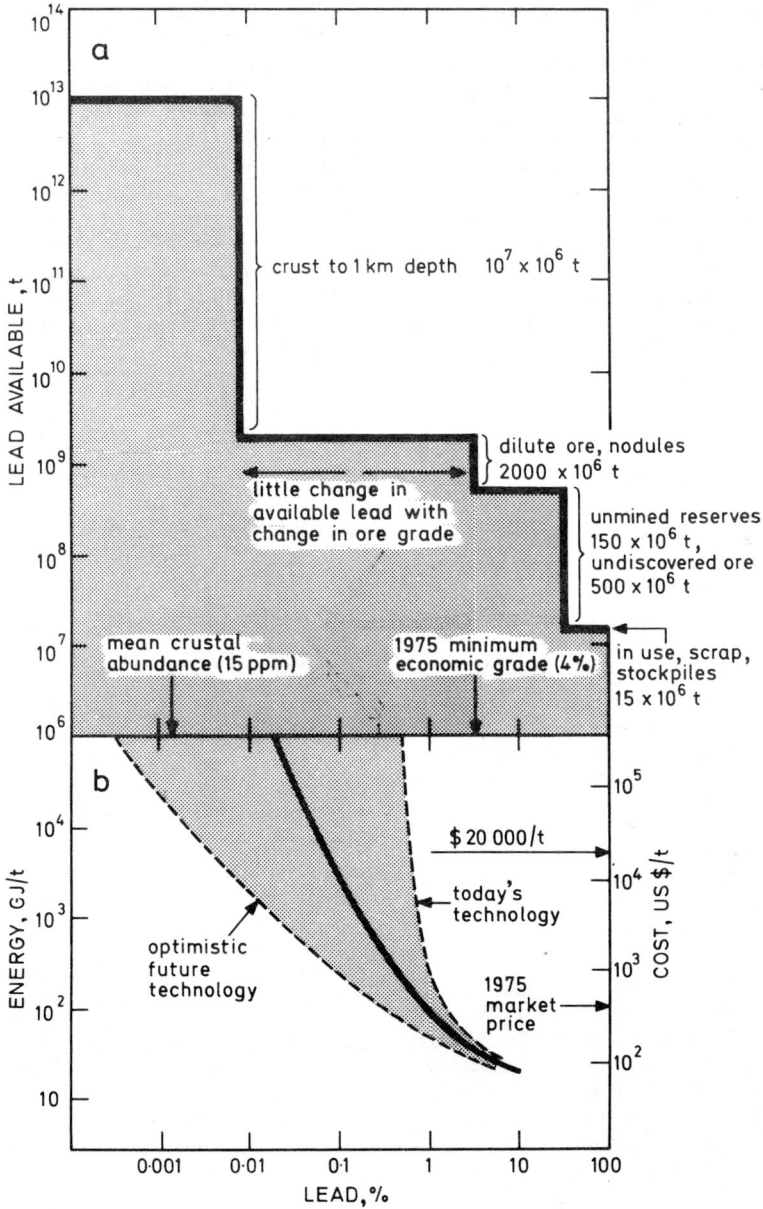

a total available lead vs. ore concentration, shown as histogram using average values; ordinate shows estimated tonnes available at concentration ⩾ those shown on abscissa

b energy cost per tonne of lead vs. ore concentration; left ordinate: energy required in GJ per tonne of lead metal (1 GJ = 278 kWh); right ordinate: cost of energy required in US dollars per tonne of lead metal

1 Availability and energy cost of lead vs. ore grade (after Ashby[11])

come in, an area in which many options are open to us.

One of the difficulties is that the world's energy reserves and resources are irregularly distributed, increasing the possibility of mismanagement by Governments. Similarly, raw materials are very irregularly distributed over the countries of the world, many of which have no significant mineral resources or reserves at all. These facts have led to the conclusion that there must be a pluralistic approach to the solution of the problem of energy/raw materials/resources exploitation in different countries. Furthermore, one can envisage a substantial movement of demand between one source of energy and another as exploitation and prices vary, e.g. Canada has already twice switched away from and back to natural gas in the last five years.

Metals and plastics

Unfortunately, probably nine to twelve of the metals we are now using will decline in availability and become prohibitively expensive. These will probably be: antimony, cadmium, cobalt, copper, lead, mercury, platinum, silver, tin, tungsten, and zinc.[11] This list is generally agreed by most world authorities who have studied these metals. What is not so readily agreed and is certainly more difficult to estimate is the time span of the half-life. Again, it could be another 30-50 years for some of them. Fortunately, the remaining metals and non-metallic materials are much more abundant and need not worry us as far as availability is concerned. What will be a very significant controlling factor will be energy resources, which we must assume will also have to be rationed or apportioned about the year 2000 and beyond. In many countries this could well be earlier but, fortunately, in the United Kingdom we may be energy-fat until beyond that date unless we have to export our oil and coal to trade for materials imports. The fact that oil is both the feedstock and main energy source for most plastics means that their prices and entire economics in relation to many metals and other materials will undergo a radical change.

LONGER LIFE AND BETTER RECYCLING ARE ESSENTIAL

One immediate and relatively easy way to conserve energy is to aim to double the lives of all products and then double them again. There is no doubt that many current metals and materials will slowly level out in growth owing to a variety of circumstances. The reasons are the lower grades of ores that remain available and the increasing cost of energy to mine and extract these ores. The same problem will also confront plastics, largely because oil is both the energy source and the feedstock.

Other materials which are much more readily available will widen their applications, a continuation of the competition for usage which has been intensifying over the past two hundred years. In such a steady-state situation, i.e. no growth for many single materials, the availability and ease of recycling old scrap becomes vital and it is incumbent upon us to design for ease of dismantling, identification, and recycling.

Unfortunately, despite the relatively low additional energy required, the world's performance in recycling old or used scrap as distinct from new or reprocessed scrap is poor.[11] For many metals which are likely to be in short supply, such as cadmium, cobalt, and tungsten, the dissipation rate is as high as 90%. But even for metals such as copper the quantity of old scrap recycled is only about 20% of any current year's production. The best performance is for lead and antimony at about 30%. For plastics it is virtually nil, and recycling of such old or used scrap back to its original properties is extremely unlikely without pyrolysis, which is bound to be expensive and of low yield.

SUMMARY

In the short term major savings in energy can be made by improving processing efficiency, concentrating on those materials which can be recycled from old scrap without deterioration in properties and also those of low total energy content such as timber, concrete, and steels. In the longer term, however, the total energy content per unit of property coupled with the cost per unit of property should be used to point the way to efficient utilization of our overall resources, i.e. energy and materials.

There will be future shortfalls in certain key materials, and we should therefore exploit every material to its optimum level in total energy terms. In the light of such an overall review of the tonnage metals and materials of the world and of the development of specialist metals, materials, and composities, it seems likely that mankind could probably manage without some of the resources which are likely to be in short supply or which demand excessive total energy. The reasons are that there is an abundance of some materials, and many properties which are desirable overlap between them. Furthermore, the ingenuity of engineers and scientists ensures that there is always more than one way to achieve a desirable end, whether it is to produce a structure or an instrument, even if at a relatively high cost.

With foresight, and if sufficient facts are made available to the world community, we need not be too downcast by the predictions of the prophets of doom.

REFERENCES

1 Energy Audit Series No. 1, 'Iron castings industry', Department of Energy and Department of Industry

2 BNF Metals Technology Centre, private communication

3 J. C. HEWGILL: Presentation I, 'Evaluation of energy use, now and tomorrow', Watt Committee on Energy, Consultative Council, 22 May 1979

4 S. S. W. LOM: 'Feasibility study on energy accounting', thesis, Univ. Stirling, 1977, pp. 25, 41

5 Ibid., p. 27

6 Noranda Metals Industries Ltd, Montreal, Canada, private communication

7 'Materials aspects of world energy needs', Int. Materials Congress, Reston, Virginia, USA, March 1979, verbal discussion

8 Int. Federation of Institutions of Advanced Study, Sweden, 1974

9 'Energy accounting of materials, products, processes and services', 9th Int. TNO Conference, Rotterdam, February 1976

10 H. H. KELLOGG: 'Conservation and metallurgical process design', Inst. Mining and Metallurgy, April, 1977

11 'Rational use of potentially scarce metals', Report of NATO Science Committee Study Group, p. 16

12 New issues late 1979 of Energy Audit Series, Department of Energy and Department of Industry

13 W. O. ALEXANDER AND P. M. APPOO: 'Material selection: the total concept', Design Engineering, November 1977

D. A. Dyker
Distortion of economic factors due to defence strategy

'Distortion' is a loaded term and 'strategy' an open-ended one, and no apology need be made for giving the military element in the economic situations of both the East and the West the broadest treatment. The first and most basic question that arises in connection with defence expenditures is: what proportion of the national income do they form? More or less precise figures on this proportion are available, and Table 1 gives selected years for the USA.

Defence expenditures do not represent a highly significant proportion of the national income for either of the two superpowers and do not exhibit overwhelming importance in total government expenditure in any of the Western industrial countries. Arguments about the indirect economic importance of defence have to be taken seriously, but it has not been conclusively proved that there is a military/industrial complex. In relation to resource utilization it is noted that in certain respects technical advance in weapons manufacture may have contributed to materials economy. In respect to energy, however, it seems clear that defence is, and is likely to remain, a most expensive business.

The author is with the School of European Studies, University of Sussex.

In America, as in other Western countries, the proportion varies enormously, not only for military reasons. Considerations of general economic policy and medium-term political circumstances have surely — though it is difficult to test such a hypothesis — often had a substantial effect. There can clearly be no argument, however, that in any but the exceptional years of the Second World War did military expenditures represent a dominating element in national income *per se*. (To put the matter into perspective, one might consider the fact that in most of the industrial economies of the world the proportion of national income devoted to fixed investment is in the region of 20–30%). The British figures suggest a similar conclusion. In recent years the proportion has been in the region of 4–5%, though during the Second World War it was over 60%.

For the Soviet Union estimates vary widely, though all of them indicate a greater degree of stability in the proportion than is usual in the West. One could speculate on the relative importance in this context of the comparative invulnerability of Soviet-type economic systems to macroeconomic stabilization problems, and the insensitivity of the Soviet political system to the kind of political pressures that built up, for example, in the USA in the post-Vietnam period. The current fashionable figure for defence expen-

Table 1   US defence expenditure as percentage of GNP

1940	1944	1953	1957	1967	1971	1975	1976	1977
1	46	15	11	9·4	7·1	6·0	5·4	5·3

Sources: various publications; figures for recent years taken from SIPRI Yearbook 1979, pp. 36-37 (1979, London)

ditures as a percentage of Soviet national income is 'around 11–12%', but expert opinion is disinclined to place too much reliance on any estimate of the Soviet proportion, and not just because official Soviet statistics are designed to conceal rather than reveal data on defence. In an economic system where prices are administratively fixed there is every possibility that defence goods may be costed far below their real value in terms of scarce resources. But however much stress we lay on this problem, there is still clearly no basis for arguing that defence expenditure in the Soviet Union forms a predominant proportion of national income.

Defence spending does of course bulk larger in aggregate government expenditure in Western countries, but in an age when government expenditure in itself appropriates around 50% of national income in most of the developed industrial economies, the defence component still appears fairly modest in most cases. Even the USA, with its lower-than-average proportion of government expenditure to GNP, does not report figures indicating any overwhelming predominance of defence in total government expenditure.

Given all this, what can be made of statements like that made by Cook, an authority on the US defence sector, to the effect that the bulk of production and supply of services in the American economy is directly or indirectly dependent on defence?[1] Justification of this thesis must clearly place primary emphasis on the factor of indirect dependence, to which we now turn.

Is there such a thing as a military/industrial complex? Certainly not in the Soviet Union, where the professional soldiers and industrialists are kept firmly in their places by a civilian leadership which is, admittedly, fairly militarily minded and very industrially minded. But we have to take seriously the proposition that in the USA there is a network of business/government relationships which makes government dependent on big business, big business dependent on government, and both dependent on defence. If we break the argument down into its constituent parts, however, we will see that it is anything but conclusive. It is true that the American defence industry is dominated by the industrial giants, but the industrial giants are not dominated by defence contracts.

'. . . Three-quarters of the largest defence contractors in America were among the top 500 American corporations in 1968 . . . [but] only a small minority of [these] corporations are by any standard dependent on defence work for their business.'[2] Despite the size of the firms involved there is rapid entry to and exit from the defence industry. The very high rate of technical progress in the industry introduces a dimension of competition normally absent in oligopolistic conditions. In 1960 thirteen firms entered the missile business and five left it. In the same year eight firms entered the aircraft/rocket engine business and seven left it.[3] The picture of huge corporations heavily dependent on defence contracts and dominating the national economy decade after decade is, then, at the very least seriously overdrawn.

Looking at the issue from the government point of view, it is incontrovertible that any element of expenditure that can be autonomously varied by the government of a market economy may be of great importance in maintaining macro-economic balance. Of course, investment is a much bigger item than defence, and policy on investment expenditues therefore *ceteris paribus* a much more powerful tool, but it is perfectly plausible that in certain circumstances a government may feel that it 'needs' to increase defence expenditure substantially in a time of

31

weak aggregate demand, or to decrease it when demand is excessively buoyant. But it should be borne in mind that the last thing that any defence lobby in industry is going to want is a level of defence expenditure that goes up and down like a yo-yo. (As noted above, American defence expenditure has indeed varied tremendously in relation to national income from one period to another.) In any case, a large proportion of defence expenditure is not on weapons — in the UK in the period 1973-74 50% of defence expenditures were on personnel. The transition from the argument that defence is important for reasons other than defence to the argument that there is a military/industrial *complex* is, then, an awkward one. But there is a further argument which even calls into question the potential for macro-economic balancing that may be offered by variations in expenditures on military hardware. It has been suggested that as the defence industry has become more concentrated, firm-wise and regionally, so the scope for wide-ranging multiplier effects has become progressively weakened. This is an aspect of the general argument that Keynesian macro-economic mechanisms are being progressively neutralized by the development of 'meso-economic' giants. Similar types of argument are often found in harness with radical interpretations of the military/industrial complex. Once again, however, it must be emphasized that you cannot have the argument both ways.

In all, then, the purely economic arguments do not point to any overwhelming predominance of the military element in the industrial economies of either the East or the West. Yet in some ways this conclusion goes against the logic of everyday observation. However controversial the issue may be in relation to the USA, there can be no question that the Soviet Union is in a very real sense a militarized society. We could of course broaden the issue here to bring in, quite properly, a variety of purely political and sociological factors. The theme of the conference does, however, suggest another, more specific line of enquiry. In an age when economists are increasingly coming to consider that materials have to be ranked as a factor of production on a par with labour and capital, is the general economic importance of defence primarily in relation to materials utilization?

To answer this question would require a large-scale research project in itself, and all that can be done here is to point to some of the key factors which would condition any answer to it. Military hardware is certainly much less 'heavy', in the crude sense, than it used to be. Electronics accounted for just 16% of the cost of a Second World War plane, and for as much as 50% of that of a 1960 plane. Recent developments in Western defence technology have surely not reversed this trend. As anyone who has ever looked at Soviet consumer goods will know, the Soviet economic system is notorious for inducing excessive 'heaviness' — quite simply because its system of obligatory output targets encourages maximum absorption of inputs in order to push up reported output. But there is no evidence that this factor is extensively present in the defence sector, and Western research has rather highlighted the extent to which the intense degree of involvement of the central political authorities in the defence industry has succeeded in creating an enclave within the Soviet economy which demonstrates few of the *systematic* weaknesses of that economy.[4] Six new Soviet aircraft recently unveiled suggest no technology gap whatsoever.[5] It seems incontrovertible, then, that a given level of defence potential uses up less steel etc. than it used to. (We are not of course saying that, for example, the USA devotes a smaller *absolute* volume of steel to defence needs than it used to.) In relation to the rarer metals, for which weight is of course a quite minor variable, the trend has obviously been in the opposite direction. The delegates at this conference are probably more aware of the details of this trend than the author can be, and may know of interesting cases where advances in defence technology have actually produced a saving in utilization of rare metals.

But surely the most important respect in which defence has continued to become more and more resource-intensive is in terms of energy. Over the last decade or so, and particularly since the oil crisis of the early seventies, measures have been taken to reduce fuel consumption in the armed forces. But while some of these, e.g. improvements in aircraft engine design, have been of a technological nature, some, such as restrictions on air training flights, are purely bureaucratic and of dubious real significance. There has been no revolutionary breakthrough which would permit the mounting of a given level of conventional defence capability on a radically reduced fuel budget, and there can be no reasonable expectation that such a revolution will occur in the near future. It may be in relation to the energy balance that defence postures affect the general economic situation most profoundly, and it is surely not unreasonable to speculate that at some time in the future the opportunity cost of conventional arms in terms of energy might become prohibitive. Does this, ironically, mean that just as we may have to rely increasingly on nuclear processes to generate power so we should consider relying increasingly on nuclear weapons as a way of saving energy in defence? It is not, of course, for a humble economist to even try to answer such a question.

Defence may, then, increasingly become the major stumbling-block to an integrated policy for energy economy, and a similar argument may apply in relation to metals utilization. The counter-argument to this is that defence needs have always provided a major stimulus to technological progress and hence to economic growth, and that there is no reason why it should not continue to do so. This argument has to be handled with some care. There are famous examples (such as radar) from the West but in the Soviet Union, as already noted, the defence industry tends to operate as an isolated enclave within the economy, and technology spillover effects from the defence industry to the economy as a whole have been shown to be very weak.[6] Even if we grant that the argument has some historical validity, we can still question how valid it will be in the future. The point is that the world economy needs not just any form of technical progress but rather quite specific forms relating to resource economy. To underline the earlier argument, this appears to be the form of technical progress least likely to emanate from the defence industry.

REFERENCES

1 F. J. COOK: 'The warfare state', 175-177, 1962, London, Jonathan Cape
2 G. KENNEDY: 'The economics of defence', 114, 1975, London, Faber
3 *Ibid.*, p.124, quoting M. J. Peck and F. M. Scherer in 'The weapons acquisition process: an economic analysis', 1962, Harvard
4 R. W. CAMPBELL: 'Management spillovers from Soviet space and military programmes', in *Soviet Studies*, April, 1972. Another authoritative source points out that specific technological weaknesses in the Soviet economy have sometimes affected defence development. For example, backwardness in welding techniques has constrained tank development. See ed. R.Amann *et al.*: 'The technological level of Soviet industry', 489, 1977, New Haven and London, Yale UP
5 'The Soviet empire: pressures and strains', to be published by the Institute for the Study of Conflict, London
6 R. W. CAMPBELL: *op. cit.*

A. A. Nijkerk
Economics of metals in view of their scarcity

The value of metal is determined by its scarcity and the cost of producing it. In this respect, it is no different from any other commodity. But since metal, unlike various other raw materials, is practically endlessly recyclable, the already vast and growing amount of reclaimed metal strongly influences

Exponential increase of metal consumption affects its price at a pace which must cause concern. As metals thus become scarcer, this trend is inflated by the appalling waste of metals thrown away, or lost for reclamation, owing to the ever-increasing complexity of consumer products. Substitution and lengthening the lifetime of metal-containing products can relieve this situation, but their effect is slight. Maximum emphasis should be laid on reclamation and recycling, for these are the best alternatives to exhaustion. They yield considerable savings in energy and they free the environment from ugly rejects. The influence of recycling on the price structure of metals has not been properly recognized in the past.

The author is managing director of Metrec Metals Recycling BV, Middelburg, The Netherlands.

its price. Many metals would not be available nowadays if their widespread reclamation had not been established practice.

The cost of manufacturing is also heavily dependent on the energy employed, and here again there is a striking difference between primary metal and reclaimed material. The concern about future availability of metals, and consequently their prices, is exacerbated by contemplation of the consumption explosion in the underdeveloped countries — which may occur in the very near future.

It was the Club of Rome, in the 1972 report 'The limits to growth', which raised before us the spectre of runaway demand and the impending exhaustion of metals. But there are many ways to avert disaster: more economical usage; new sources to be discovered thanks to advanced technical methods of (re) producing metal, even extra-terrestrially; lengthening of lifetime; substitution; and, possibly most important, recycling. These factors will be discussed in turn.

SCARCITY: THE RIDERS OF THE APOCALYPSE

Scarcity has many faces, many causes. It will occur when production cannot keep pace with demand, in which case it may be temporary. It can also happen that a raw material becomes exhausted — usually because it has been used carelessly. The first Rider of the Apocalypse of Scarcity is probably the exponential* consumption of metal — exponential because there will not only be more people to consume metal but many of these people will demand more metal per head. Take for instance copper consumption: worldwide it is about 2·2 kg annually per head, while in China it is not more than 0·4 kg; whereas countries like the UK (9·1 kg) and the USA (9·2 kg) are well above the average. It is obvious that the inhabitants of the People's Republic of China are only too eager to attain something approaching the same usage of copper for their daily needs as the countries of the Western World. These 900 million Chinese alone would, once they have arrived at the UK consumption of over 9 kg per head per year, monopolize the entire world copper production of some 8 million tons. So scarcity of copper has to be feared, and for other metals such as nickel it will not be different. Any notion that we might go to the centre of the earth to get the billions of tons of nickel out of the central core (about 6 500 km in diameter) of Mother Earth is a fantasy worthy of Jules Verne himself. In practical terms, digging deeper means that we have to

*A dodge for calculating the doubling of consumption: divide the number 70 by the growth percentage. Example: a 5% growth rate will double in 14 years

Table 1 Copper, lead, and zinc in use in the USA 1940–72 (estimated, 10^3 short tons)

Year	Copper*	Lead†	Zinc†
1940	14 735	180	264
1945	19 933	873	1 173
1950	24 169	1 674	1 722
1955	28 615	2 490	2 525
1960	32 630	3 052	3 085
1965	37 346	3 587	3 928
1968	40 333	4 030	4 477
1969	41 447	4 137	4 677
1970	42 509	4 234	4 841
1971	43 615	4 405	5 025
1972	44 927	4 701	5 234

US Bureau of Mines, Office of Mineral Resource
* 1907 reference base
† 1939 reference base
  1972: evaluation

Note: world production of these metals in 1972 was approximately 8084/4071/5541 kt respectively

content ourselves with poorer ores, with more tailings and mining wastes, and all of these factors demand ever-growing energy. The initiatives taken in the face of imminent depletion often falter owing to sharp rises in energy costs, not only in general but especially wherever the metal content in the ore becomes poorer.

Table 2 The Rhine as sewer

Amounts of waste metal brought into the Netherlands from Germany, France, and Switzerland discharging their waste metals into the Rhine (I) in 1973 (in tons).
By way of comparison the total amount of 'imported' (II) and locally (III) discharged waste metal is also given.

Metal	I via Rhine only	II via Rhine + Meuse + Scheldt	III discharged by local polluters
Copper	1 650	2 000	50
Zinc	13 000	15 000 – 20 000	1 500
Lead	1 700	2 000 – 3 000	200 – 300
Chromium	*	3 000 – 4 000	200 – 300
Mercury	40	80	4
Arsenic	400	400 – 500	15 – 20
Nickel	*	800 – 900	20

*Unknown

Table 3   Lead in petrol (10^3 tonnes)

	1970	1973	1974	1975	1980	1985
USA	253	245	277	180	105	85
W. Germany	5	9	12	9	5	5
UK	40	54	56	55	50	45
France	12	14	14	12	12	10
Italy	15	12	10	9	13	10
Others	30	32	28	25	25	20
TOTAL	355	370	347	290	210	175

If energy, in this respect, is possibly the second Rider of the Apocalypse, the huge amount of metals lost is then the third. Here is where we need to be re-educated in order to avoid the catastrophe of total depletion. We are rather careless with metals (except when a war brings us temporarily back to reality); we consume them and throw them away. According to Lavoisier nothing will be lost on this earth, certainly not the pure elements that most metals are. True, the loss of metal in a rocket fired into the dark universe is not dramatic; indeed, it may well be off-set by the gain from meteorites which have provided us with so much (even pure) metal during the course of history! But there are other factors. Notwithstanding Lavoisier many metals are lost, sometimes temporarily, sometimes in practice for ages or forever.

As the price of copper rises, the four bronze propellers of the *Lusitania*, lying at a depth of 315 feet off the coast of Ireland, become a more attractive prize for divers. Perhaps one day we shall be able to raise the 40 000 and more shipwrecks lying on the bottom of the North Sea. But what about the rust blown away by the wind; the hundreds of thousands of tons of lead pumped into the atmosphere by the exhaust pipes of automobiles? What about the zinc used in galvanizing (the main reason why only about 25% of all zinc used is reclaimed); what about the 150 000 tons of copper consumed annually in fungicides, especially in potato growing; the silver lost in photography — over one-third of annual production? The explosive rise in the price of tin is only partly due to politics, inflation, or rising energy costs; the main reason is scarcity, caused by careless use. Cans made from tinned steel use up the bulk of tin production, but only a minor percentage of them is recovered, while the tin itself is mainly lost. Incinerators, or simple atmospheric rusting, cause the disappearance of tin, but in addition more than 10 000 tons of tin are lost in chemicals: fungicides, stabilizers, the solders used to seam a tin can.

SCARCITY AND THE ALTERNATIVES

Of course, we can be re-educated to use our metals more carefully, to look for alternatives to simply throwing our metal in the garbage bin. Separate collection is certainly one (small) step towards a more economical use of metal, and any effort to economize on the weight of consumer products is another step. New techniques also provide us with alternatives. They will allow us to dig deeper, to get a better yield from ores, and to recover all its by-products. Cadmium and silver are recovered almost completely from zinc ore, for instance. There are even new sources in sight which were not thought of only 20 years ago. There are the manganese nodules lying some 15 000 feet deep on the bottom of the Pacific; these will yield a huge reserve of nickel and copper once they are literally sucked off the ocean bed by sophisticated 'vacuum cleaners' able to endure the pressure at such great depths.

Magnesium is produced out of the sea around us, but seawater contains nearly every metal. The present-day price of copper does not invite the boiling of seawater, as it would take 300×10^9 litres of water to generate one ton of copper. And it is still not economical to make aluminium out of ordinary garden clay, even though one-quarter of it is aluminium, because it is tightly combined with silicon, calcium, magnesium, and oxygen. But the time may come ...

Lengthening the lifetime of consumer products would undoubtedly lead to a lower consumption of metal and so postpone its exhaustion and mitigate price rises. But extension of the life span has its drawbacks as well. Recycling time span will then also be delayed and, as metals are practically endlessly recyclable, the circle will be closed at a slower pace. On top of this, many consumer articles are meant to have a short life. Would it make sense to make tin cans which last a hundred years, or durable aluminium wrapping foil? This prolonging of the life span will also make consumer products more expensive; to make a Rolls-Royce from a Mini is possible,

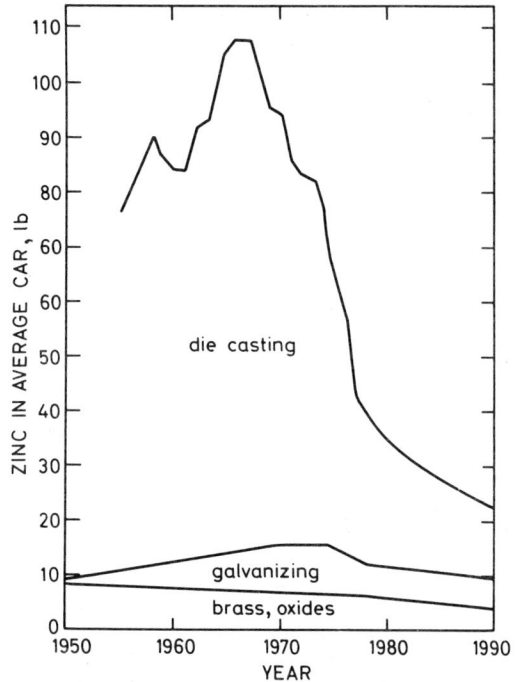

1   Zinc in motorcar construction (USA)

Table 4   Seawater: resource of the future

Seawater covers over 70% of the earth's surface with a volume of $1450 \times 10^{15}\,m^3$

Seawater contains 3·5% of various salts, i.e., 35 million tons of chemical compounds, notably (g/litre):

magnesium sulphate	2·34
magnesium chloride	3·58
magnesium bromide	0·05
kalium chloride	0·72
natrium chloride	27·11

and of metals:

Sb	0·05
Bi	0·002
Cd	0·01
Cr	0·005
Co	0·0004
Cu	0·3
Hg	0·003
Pb	0·003
Mn	2·0
Mo	1·0
Ni	0·2
Sn	0·08
Ti	1·0
U	0·3
Ag	0·004
Zn	1·0

but such a car would devastate one's bank account. A nickel–cadmium battery has a longer life span than its lead–antimony brother, but the price difference is still great. Consumer behaviour and taste have also to be taken into consideration: what lady would wear a dress for 20 years?

IS SUBSTITUTION AN ALTERNATIVE?

Substitution for metals can be made in two ways: we can replace a scarce metal with one more plentiful, or we can divert away from metal towards any non-metal such as plastics, wood, glass, etc. Here, economics will find its own way. When a metal becomes scarce it becomes more expensive and consequently we begin to search for alternatives. Substitution within the world of metals is daily news. It is

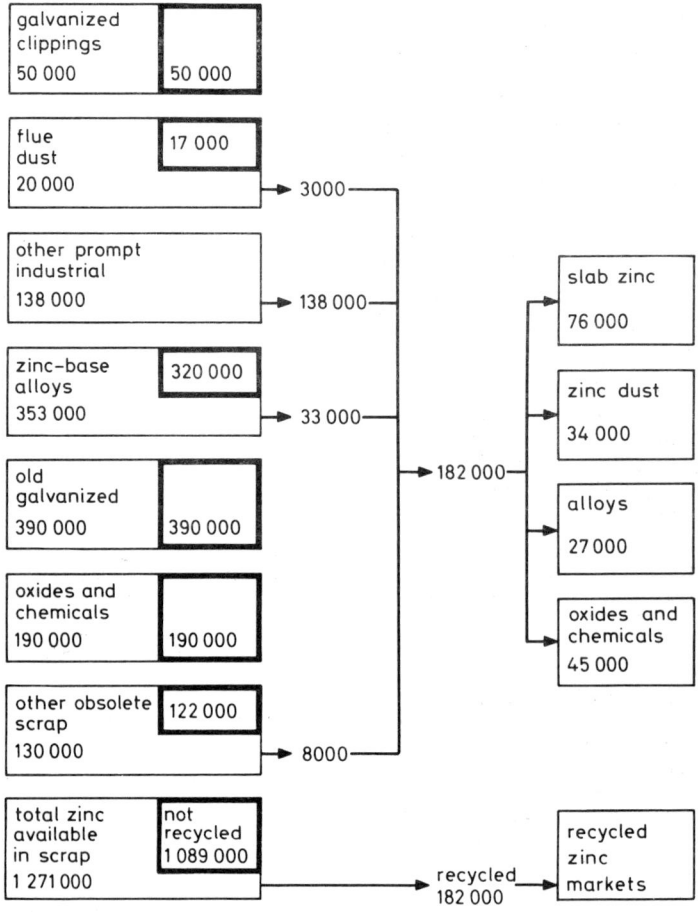

Notes: all quantities in short tons of zinc; zinc recycled as brass is not included

2   Flow of scrap and recycled zinc

Table 5   Secondary production as major proportion of raw material supply

Scrap	Percentage of total US consumption
Aluminium	30
Copper and brass	45
Lead	52
Zinc	20
Paper stock	20

a constant struggle between producers of one metal and another. When electrolytically tinned steel sheet was failing to save enough tin compared with hot-dipped tinplate, then the aluminium can emerged, and even the TFS (tin-free steel) can. Calcium may substitute for the antimony in car battery lead, and the weakness of zinc prices is partly due to using aluminium as a substitute for zinc in the zinc alloys (or zamac) in cars and, to some extent, plastics. The war has taught us that aluminium, reinforced by a steel core, can replace copper high-tension overhead wires, and many examples can be found of one metal, often aluminium, taking the place of another, usually copper or zinc.

APPLICATIONS	Al	Pb	Cu	Zn	Sn	Fe	plastics	various
COND. MATERIALS, BATTERIES cond. materials cable sheathing								
structural building								
tubing								
printing								
MOTOR VEHICLES engines								
ornaments,								
interiors								
bodies								
railway carriages								
BUILDING MATERIALS prefabricated roof, front and other architect. building elements								
PACKAGING								
foils								
collapsible tubes								
cans								

EXAMPLES
1. Al ⟵──#Cu   aluminium replaces copper
2. Al ⟵──── # aluminium partly replaces tinplate

3   Main areas of substitution

Substitution has an educational effect as well, as it forces an industry using an endangered metal to look for cheaper ways of producing or using it.

However, substitution often goes much slower than expected or predicted. True, plastics have replaced metals in many consumer articles, but we have also learned to appreciate the manifold properties of metal. Why have we not yet seen a breakthrough — as predicted decades ago — leading to the all-plastic car body? Because steel, as we have come to realise, has a combination of properties which cannot easily be matched. Steel is safe, as it has great impact strength, folding in the event of a crash; it is fireproof, easy to repair, and to cover with a lacquer. Plastics fail to provide such a combination of good properties.

Some examples of substitution practice are shown in Fig. 3.

RECYCLING –
THE BEST WEAPON

The Three E's: Exhaustion, Energy, and Environment all act to the detriment of primary metals and in favour of recycling or the secondary metal. Thus the balance of supply may weigh in favour of recycling, both in tonnage and price. As to tonnage, it may entail that more tons of scrap will in future be used in production of metal than the total metal content in ores. In ferrous scrap and lead in the United States, for instance, this 50% boundary has already been passed, and other metals may follow suit. A copious collection of scrap will influence the price as well; that is, scrap may become relatively cheaper, especially as energy costs continue to rise rapidly.

Less energy is required to remelt secondary metal out of scrap than primary metal out of ore. The costs of collection also differ: the energy required to gather scrap ('mining above ground') is far less than the total energy requirement for digging ores and transporting them to the smelter, often after concentration on the spot. The price mechanism of metals will therefore be more dependent on the availability (reclamation) of secondary metal. Secondary zinc and aluminium, which are up to now only 20–30% reclaimed, will have less

Table 6   Energy requirements and savings:
melting of primary ore vs. recycled metals

	Production (approx.)		Savings (approx.)	
	Ore, kWh/t	Recycled metal, kWh/t	Recycled metal, kWh/t	Savings as oil (barrels)
Aluminium	51 000	2 000	49 000	29
Copper	13 500	1 700	11 800	7·7
Iron and steel	4 000	1 500	2 500	1·5
Magnesium	91 000	1 900	89 100	52
Titanium	126 000	52 000	74 000	43

influence on the total metal price structure than lead and copper, which are recycled to a much higher extent. Price fluctuations have a practically immediate impact on the collection of secondary metals (another aspect of their difference from ores) as, for instance, the year 1974 showed; suddenly, significant amounts of metal which had formerly been considered obsolete were reclaimed. Millions of old abandoned cars rusting idle and abandoned along the US highways were now sought after and brought to the shredders; every piece of copper, nickel, and so on was considered for its recycling value. Examples of silver, nickel, and copper coins which had a higher recycling value than their monetary face value were commonplace. Old electricity cables were dug out of the earth and housewives became more inclined to embark on separating their domestic refuse — metal, glass, paper and the rest — than before.

How striking this 'hausse' effect is may be seen by looking at the example of ferrous scrap collection in the Netherlands. A slow and expected increase of some 5–10% per annum was suddenly interrupted when it exploded in 1973-74:

TOTAL COLLECTION (Mt)
1971   1·2
1972   1·35
1973   1·60
1974   1·85 *(price explosion in first half)*
1975   1·3
1976   1·35
1977   1·40
1978   1·70 *(price expl. at year's end)*

Probably 1979 will show another upheaval in collection. Scrap prices are very flexible and often anticipate a 'hausse' or 'baisse'; this is logical where ore contracts are often long-term agreements, while scrap is spot business.

Table 7   Resource recovery rates

Material	Available for recycling, short tons	Recycled, short tons	Percentage recycled
Aluminium	2 215 000	1 056 000	48
Copper	2 456 000	1 489 000	61
Lead	1 406 000	585 000	42
Zinc	1 271 000	182 000	14
Nickel	106 000	42 100	40
Steel	141 000 000	36 700 000	26
Stainless steel	429 000	378 000	88
Precious metals (troy oz.)	105 000 000	79 000 000	75
Paper	46 800 000	11 400 000	19
Textiles	4 700 000	800 000	17

POLITICS

Politics play another role. There is a visible flow of primary metal in the form of ores and concentrates from South to North, from the developing countries to the highly industrialized world, where the centres (USA, EEC, USSR, Japan) all lie in the northern hemisphere. The industrialized countries have a traditional demand for ores and their scrap collection is well developed as there is a great historical arsenal of metal which can be recovered. The developing countries may well have ores but lack the money to build factories to work on them and, besides, have a relatively low reclamation level; so that, if they aim to keep pace in industrialization, they must either import scrap to build up a comparatively cheap secondary industry, or borrow money to build primary industries. In the latter case the developed countries have to secure their future ore supply or else pay more attention to reclamation. Thus in the year 2000 we may conceivably see a quite different North/South division: primary metal produced in the Southern World, secondary in the Northern.

Economists can no longer ignore or discount the role of secondary metal when building their models of future demand and supply. In this respect the report of the Club of Rome needs to be rewritten, for it did not give adequate consideration to the role of secondary metals. Like everything else, even a report can be recycled!

T. J. Tarring
Markets

Whereas the topic of every other speaker at this conference is either a specific metal or metals or one to which metals are central, my subject, 'Markets', is not really about metals at all but about people. There are, after all, markets in anything one cares to name; not only is there an international market in every commodity and raw material but there are markets in finished products and even in their end products. Commentators on metal markets are obliged to refer to the state of the motorcar market, the building market, the capital goods market. The one thing that is common to all these markets is *people*.

If people are reluctant to put their hand in their pocket to buy a bar of chocolate, that is bad for the cocoa market; if they

---

Like any other market, that of the metals industry has to be related closely to the workings of human psychology, which changes little over the millennia. The pricing practices in physical trading are also innately resistant to changes which could be made to help to adjust production and consumption more readily during the next decades to future needs and resources.

The author is managing director of Metal Bulletin Ltd, London.

do not want to buy shirts, that affects the textile market; and if they decide not to buy a car, that depresses a number of metal markets in the near term and is also bad for the capital goods market in the longer term. Conversely, of course, if they do want to buy it seems that nothing can stop them. Certainly the operation of the price mechanism under the age-old law of supply and demand does not inhibit their urge to buy.

Indeed, for the very reason that the constant factor for people is their perverseness, one can virtually guarantee that as soon as prices start to rise they will redouble their efforts to buy until prevented by actual non-availability — the condition otherwise known as a 'shortage'. It is interesting that in a discussion such as the present conference shortages are viewed as a bad thing; they spontaneously attract the pejorative adjectives. On the other hand, in the vocabulary of markets it is rising prices (in other words, emerging shortages) that attract the favourable adjectives. For example, while a cablemaker is talking of a 'disastrous shortage', his broker on the London Metal Exchange is calling the market 'buoyant', or in even greater conditions of shortage 'euphoric'. This should tell us something about the way in which markets are firmly rooted in human psychology.

It follows that if there are any radical changes in markets which we could discuss in this meeting, they would have to be related to human psychology rather than to any metallic consideration. Unfortunately for the purposes of our discussion, people are exceedingly slow to change. This meeting has set its forward horizon just 21 years ahead; but I can assure you that in its fundamental responses to outside stimuli the human race will hardly change at all in the next 2000–5000 years, just as it has changed little during the last 5000 years (easily proved by looking anywhere in the Bible from Exodus to the Acts of the Apostles — the early Jews were certainly fully alive to the law of supply and demand).

This leads me to a constructive point about markets which I feel can usefully be made in the context of this conference, since we are here to discuss future metal strategy on the basis that we can do something to influence the future course of events that we believe we can discern. The position within the market at which I believe the industry has it in its power to act is at the interface between the source of prices and their application.

It will, I hope, be readily agreed that for copper the London Metal Exchange is a price-making institution while the industry, through its annual contracts between mine producers and first-line consumers as well as in its application of LME prices to other areas of pricing, is a price-taking body.

The LME closely relates to what I said earlier about the universality and immutability of human nature and it would be Canute-like for me to suggest changing the LME. This is not to say that the Exchange is incapable of change: on the contrary, its slow steps towards the introduction of its so-called monitoring scheme constitute probably the most fundamental change in its approach to self-policing since the Exchange was founded over 100 years ago. Indeed, it is one of the remarkable aspects of the worldwide influence which the Exchange has gained over the years, especially in copper, lead and tin, that so much trust has been placed in the conditioned reflexes of so few organizations with so little by way of a formal financial framework to their relations both with one another and with their clients. Let me hasten to add that one cannot deny the argument put forward over the past three years by the opponents of

a change to monitoring: 'Nothing's gone wrong so far'. Yet those same words could have been used by the world nuclear power industry until the alarming episode of the Three Mile Island power station in the USA.

So it is important to realize that in going through its almost non-stop performance of bidding and dealing the Exchange itself does not consider its function as a price-maker for a large segment of the relevant industry as a prime objective. It is simply concerned with executing its clients' orders, reading the signs and portents in order to do this to best advantage. Let me make it clear, too, that when the Exchange is used as a hedging medium by that sector of the industry which is intelligent and sophisticated enough to do so, another functional need is met.

I will now say something which may startle, for it is my contention that the large part of the industry that has no functional connection with the Exchange at all but which follows its prices eagerly, even slavishly, is in fact engaging in an optional activity. There is no law compelling either producers or consumers to index their period pricing arrangements on the LME; moreover, there is certainly no pressure from the LME itself upon the industry to use its prices.

This point is proved by the response of the aluminium and nickel industries to the introduction of trading in these metals on the Exchange. In practice they have not responded at all; some traders have operated on the markets, either speculatively or technically, but for all practical purposes there has been no classic hedging by the industry nor any real volume of physical business through the Exchange. The counterpart to the trader business, therefore, is pure speculation from the commission houses.

Since we have now encountered (as was inevitable) the emotive word *speculation*, I will immediately place in context the activities of the speculator on a fundamentally trade-based futures market — which the LME is. (Here I should like to interject a word about the vast increase in the relevance of Comex to the trade in the USA following the adoption of Comex prices as a reference by some of the more influential mine producers. Despite this move, it is still broadly true to say that Comex remains an inward-looking, speculation-oriented market. This can be demonstrated by comparing actual deliveries of physical metal out of warehouse from Comex and from the LME.) As I said, the LME is a trade-based futures market on which the classic basis for client operations is to hedge an unbalanced position in physical, thereby eliminating exposure to risk in that position. Nothing is more annoying to me — and I carry no banner for the LME, I assure you — than to hear a fabricator or smelter remark in a smug and pious manner: 'Oh no, we never hedge; we would never speculate on the Exchange'. Hedging, I repeat, is a process designed to eliminate risk for a member of the trade with a physical position arising out of his normal operations which he needs to protect from price fluctuations. In other words, it is insurance; and, just as in insurance, there must be, for every person wishing to hedge a risk, someone willing to assume that risk. On the Exchange, in opposition to the trade client of the market, the risk-taker is usually the speculator. The trade hedger, the physical producer or processor who is inherently long of metal, goes short on the Exchange and so avoids a loss on his metal value in a falling market. He also makes the contango, if there is one. At such a time his opposite number has to be a speculator who is willing to buy three months in the hope that by the time it matures the market will have appreciated.

I have touched on these basic functions of the Metal Exchange because an understanding of this mechanism is essential to recognizing where, in the future, beneficial changes might occur in the way that prices emanating from the Exchange are used by the trade as a price-taker. It is very common to complain about the day-to-day volatility of Metal Exchange prices, but the basis for these complaints is simply a failure to understand that it is not the Exchange's main purpose in life to produce prices. Once it is understood that the day-to-day gyrations of the futures market are essential to its own internal workings but not essential to the march of monthly or (even better) annual average prices, we can see an opening for introducing a change in the way in which the industry uses the Exchange

as its price-maker. For just as there is no compulsion upon the industry to use the Exchange prices for reference purposes at all, so there is certainly nothing to stop it reading those prices off in the way which suits it best. In terms of stability a monthly, quarterly, or (ideally) a rolling annual average price offers the price-taker the best of all worlds, for it reflects the really substantial changes in supply and demand which must be read if production and consumption are to adjust themselves to new long-term realities; while at the same time it virtually eliminates the irritations of day-to-day fluctuations.

Apart from inviting you to think about this point I will not elaborate it further but use my remaining space to refer to the two other major sources of reference prices used by the industry to keep abreast of its markets without actually haggling over the price of every single delivery. These sources are, of course, (1) producer prices and (2) trade journal evaluations.

Producer prices are rather like futures market prices in that there is little the industry can do about their method of formation, which is done by the producer 'by the seat of his pants'. Unlike the LME, the consumer has little option as to the way in which the producer price acts as his reference price. In a seller's market he can either take it or leave it, and is probably happy to take it since the free market will be at a premium. In a buyer's market he keeps pushing for discounts or benefits like consignment stocks and worries about how much better his competitors may be faring. In those industries such as aluminium, nickel and molybdenum which are historically producer-priced markets it is difficult to visualize much change occurring in systems before the end of the century; quite a few steel products could be included in a somewhat broadened version of that statement.

As for those markets where a trade journal evaluation has been adopted as the source of the industry's reference price, nowadays this covers virtually all markets which are not already covered by a futures market price or a strong producer price, and thus includes many ores, as well as alloys and scrap. Here the scope for introducing change is much wider. I cannot speak for other publications but I can certainly say that for *Metal Bulletin* the process of evaluating price indications is an extremely democratic one. Not only do we welcome contributions about recent prices actually traded from any qualified source, large or small, producer or consumer, but we always encourage comment on the method of formation of the quotation concerned.

As in other democratic institutions (Westminster for instance) much of this debate is inconclusive or ebbs and flows across the same ground; but in course of time changes do occur which reflect the majority wish of the industry. Unfortunately — and here we revert to the subject of human nature — the representations we most commonly receive are from consumers who have an ingenious argument as to why the price should be adjusted (or its methods of formation changed) in a manner which suits them, closely followed by producers arguing similarly on their own behalf. For the most part these two pressures cancel each other out, but if this fails to happen we have to adopt a righteous attitude and insist that we cannot depart from our historic role of viewing these questions with complete objectivity. That is one matter about which I can make a firm forecast, for indeed we shall never depart from that role.

The discussions which followed each presentation at the conference are recorded verbatim in a separate section at the latter end of this volume.

SESSION B
The metals

S. J. Ross-Macdonald
Aluminium

Today aluminium is second only to steel as the most widely used metal; and more aluminium is used than all other non-ferrous metals combined. There is scarcely an industry, natural resource, or consumer sector of a developed economy that does not consume aluminium in sizeable quantities. This has not always been so nor, in my view, has this position been reached solely on the merits of the metal itself. Rather it reflects the attitudes and ambitions of the early proprietors and managers of the primary smelting sector as it emerged during and after the second World War into a worldwide industry. These men saw the solution to their problems in large-scale projects, based on cheap uninterruptible

---

Aluminium has achieved its present position as the second most widely used metal, with a consumption exceeding that of all other non-ferrous metals put together, by the pursuit of volume-based products and growth during the 1950s and '60s. However, this strategy has proved to be relatively unprofitable and has led directly to government shareholding in one-third of the world's 51 primary smelting companies. In addition, over half the primary metal passes forward to an integrated semi-fabricator. Its high energy dependence and the demand of enormous capital formation for new projects places aluminium smelting as a high-risk venture. Looking forward, there is ample evidence that there are sufficient bauxite reserves to take aluminium production well into the 21st century. While the possibility of scarcity might arise from political considerations in view of the location of the bauxite, it is considered unlikely that producing countries could achieve a cartel as effective as the one presently existing with crude oil. The question of price trends for smelter power supplies is very complicated. While aluminium smelting is highly dependent on electricity (preferably non-interruptible) it is felt that provided the cost of electrical power is not disproportionately higher than prices for new fuel or new petrochemical feedstock, then aluminium can maintain its position vis-à-vis its two principal competitor materials, plastics and steel. It is suggested that the next phase of development will probably see the major multinationals acting as turn-key contractors for governments rather than shareholders in their own right. In this climate the consumer is probably going to be the principal sufferer, since he will not warrant as much attention in these circumstances as he used to enjoy when he was dealing with shareholding multinationals.

The author is general manager of Bahrain Aluminium.

energy, often remote from existing or potential markets and relying on very long term returns on what were then regarded as blue chip assets. This policy was applied to a metal that required large quantities of electrical energy in the reduction of its oxide, for which the reduction process requires truly continuous operation within the life of the reduction cell and is very capital-intensive. This demanded a metal offtake which paid almost no regard to variations in the level of economic activity, the ability of fabricators to satisfy latent market applications, or the readiness of the market itself to see the need to switch to aluminium from other materials.

As with all ambitious men the solution to these problems always lay outside their bailiwick: if the smelters themselves went into semi-fabrication on a large enough scale then the technological changes could be legislated for from above and costs would come down to help market penetration. So they went into semi-fabrication. Once the unending quantities of sheet, extrusions, and castings became available it was just a short step in justification to see the necessity to purchase further downstream and make the fully finished products themselves — and this was what they did. If this was not enough, then money and resources devoted to new applications, substitution of other materials, and throwaway products would also nudge the volume of aluminium consumption upwards. So they found the money and resources. The effect was dramatic. New mills sprouted everywhere, all under the warm and comforting identity of some international group. Semi-products were made in impressive quantities and qualities and industry commodity prices for such things as can stock and oilfield drill pipe were set for long stable periods forward to pull the metal through this glittering train of investment. Everywhere aluminium was talked of; humdrum items such as buttons and cloth were studied for their consumption potential in bright laboratories as far afield as Switzerland and North Oxfordshire, while smooth graduates combed the industrial cities of the world armed with glossy brochures to find someone as yet unpersuaded of the merits of this wonder metal. However, there was a snag. Nobody made much profit. In fact, the original smelter companies made less

than before. The weight of public relations and advertising activity which had accompanied this lurch downstream had been too effective and encouraged total newcomers to the field. Governments too, lured by the glittering prizes, wanted to have this vital industry in their own countries, offering subsidies, stable power contracts, and even investment shareholding. This merely added to and compounded the chronic problem of excess capacity and low profitability, which in a climate of job protection and balance of payments deficits encouraged governments to act 'decisively', as it is known, and intervene even further.

If we now take stock of the industry we can make the following points:

1. It is relatively unprofitable
2. Of 51 primary smelting companies, 17 involve government shareholdings
3. Over half the primary metal in the Western world passes to an integrated semi-fabricator
4. The high energy dependence in the post-oil-shock world casts doubt on its future
5. New projects demand enormous capital formation and represent a high risk in the present profit record and political climate
6. Government price controls now operate in four Western European primary producing countries and in the USA.

In the period 1975-79 we have seen all these negative factors lead to a very low level of new investment, producing a balance in supply and demand if not an actual shortage. The short-term price has just reached levels where new investments from borrowed funds might be justified, though there is no evidence that this can be maintained for a sufficient period forward to service the debt involved. What is being planned now in terms of new capacity is either fairly modest expansion in the developed world or problematical, grandiose, and expensive projects in the less developed world, all involving government investment. The first is insufficient to satisfy even modest increases in demand, while the second could lead us further into the world of unreality at a time when we have shown ourselves to be ill-equipped to survive in the present all too real world.

Now for the future, or more properly the year 2000, looking at the interrelated problems of bauxite, alumina, and aluminium metal production before proceeding to certain generalizations.

BAUXITE

Total world bauxite reserves are estimated to be over 20 billion tonnes. In addition, vast deposits of low-grade bauxite and non-bauxite sources of alumina are known to exist and, significantly, some of them are in the industrialized world. It seems probable that known bauxite deposits will be adequate to supply the aluminium industry well into the 21st century. The possibility of scarcity is more likely to be a political one.

The bauxites currently mined tend to have an alumina content seldom less than 40% and sometimes as high as 55%. Since the ratios of bauxite to alumina and alumina to aluminium metal are both about 2:1, bauxite mining viewed purely as an earth-moving operation involves an effective ore grade of better than 20% compared with 5% for zinc and 0·4% for a new open-pit copper mine. With modest overburdens and less need for blasting, the physical aspects of bauxite mining involve comparatively moderate expense and capital. To produce 100 000 tonnes of copper metal from a 0·4% ore grade involves moving over 25 million tonnes of ore while only about half-a-million tonnes have to be moved to produce the same amount of aluminium metal.

It is sometimes suggested that there is no commodity better suited to imitating OPEC than bauxite. Certainly there are important similarities between them, and an International Bauxite Association was formed in 1974. Bauxite production is concentrated: Australia and Jamaica alone account for some 42% of world output and the other members bring the total to 66%. With the important exception of Australia, the main IBA producers are fairly close together in the Caribbean and North Latin America (a remarkable similarity to the Arabian Gulf and the oil world). While the IBA members are neither important producers nor consumers of primary aluminium, not one of the six largest aluminium-producing countries in the world is more than 50% self-sufficient in bauxite

and most have no significant production. The vertical integration among a few multinational companies left them with considerable scope for determining geographical patterns of trade and adjusting transfer prices to minimize global tax liability. Whether real or imagined, this power was resented among the bauxite-producing countries. As a result, the bauxite exporters have been moving towards greater participation in mining, greater local conversion to alumina, and greater tax/royalty yields. But the scope for a hoist in prices of oil-shock proportions is limited by the fact that bauxite is not the only source of alumina, although it is the most economic. Also, aluminium having gained its pre-eminence by substitution can itself be resubstituted. It is likely that the ultimate consumers of aluminium could stand a cut-back in volume for longer than the less developed bauxite producer could go without the revenues. It is uncertain therefore whether the take on bauxite as a percentage of the aluminium metal price will increase.

ALUMINA

Alumina plants in the past have been located in the major importing countries but there is now a growing economic argument for locating them at the bauxite mine. This is particularly important where the mine is a considerable distance from a deep-sea terminal, since overland freight is several times more expensive than ocean freight. As far as energy is concerned, the alumina plant can be a self-contained unit supplied by thermal power independent of public supply. However, it is important to recognize the factors of scale and regularity which, together with the development of bulk-carrying ocean freight, meant that it was cheaper to carry twice as much bauxite as alumina over the same distance and route. It was also true that nearly five times as much alumina could be carried as aluminium metal on the same route and over the same distance. There are signs that this bulk-carrying logic no longer holds and will force a bias towards transporting the minimum rather than encouraging the movement of the largest tonnage. This will favour the location of the alumina plant at the bauxite- or alumina-bearing ore source.

SMELTER

As far as the smelter is concerned, location has largely been determined by the availability of cheap, reliable electric power, generally in or near the industrialized countries where most of the consumption of the metal has been. However, the location of smelters in Bahrain, Ghana, and New Zealand was based on cheap power well away from sizeable consumption of the metal. In addition it is worth noting that cheap power is not by itself sufficient to attract a smelter. The Alba smelter in Bahrain is fired by produced gas which could be left in the ground, while 20 miles across a shallow channel enough Saudi Arabian associated gas is flared to smelt 6 million tonnes of aluminium per year. Additionally, a smelter is nearing completion in Dubai to be fired by waste gas, the supply of which is now said to be far from certain. Not 200 km away vast quantities of flared gas testify to Abu Dhabi having the highest gross national product per head in the world.

The question of price trends for power supplies is very complicated. The cost of purchased electricity was stable for many years because whatever minor inflation existed was offset by improved power station efficiencies and the scale factor economies of larger units. These beneficial effects reached their nadir ten years ago when inflation rates began to escalate and when pollution abatement, strict coal mining regulations, higher interest rates, and longer lead times became major cost elements in power production. It has been calculated that the weighted average cost of electricity to existing non-communist smelters is about 15 mills per kWh, or at today's efficiencies US$230 per tonne of metal. The range is 50–550 US$/tonne.

Where these prices are going depends on whether the power supply involved is smelter-owned hydroelectrical power, smelter-owned thermal power, power purchased from expanding hydroelectric or nuclear power systems, or purchased thermal power. Taking into consideration the present mix of power sources, the percentage of self-generation, and the nature and duration of present contracts, it is estimated that today's weighted average cost will increase annually by 70% of the general inflation rate. If this turns out to be 7% then the weighted average smelter power cost would rise by about 5% per year and

would reach US$500/tonne by the year 2000. At the same time an estimated cost for new firm power of between 95 and 105 mills per kWh or as much as US$1250/tonne is likely, some 250% greater than today's highest cost and five times the average. However, it is useful to note that today energy represents about 30% of the market price of aluminium. It happens that two principal competitive materials (plastics and steel) also have about a 30% energy content. The important difference, however, is that in the case of plastics and steel electricity accounts for only about one-tenth of the cost of their total energy requirement, while for aluminium electricity represents almost two-thirds. If prices for new power are going to be proportionately much higher than prices for new fuel or new petrochemical feedstock, then aluminium will face serious competitive cost problems.

With the single exception of the dependence of present aluminium smelting technology on so much firm electrical power, aluminium can be considered well placed to maintain its position vis-à-vis its strongest competitors—steel, plastics, timber and, to a lesser extent, copper. More research will be made into the design of a flexible reduction cell which can withstand more heat shock resulting from power outages and can operate effectively over its life with a variable power input so that it can follow the cycle of peak and off-peak power; a cell perhaps that can be switched off and restarted easily to cope with strikes, walk-outs, and just simple things like lack of water in a hydro-system. In a world where the certainty of uncertainty is increasing, flexibility can command a premium. There will also be a much stronger recognition of the energy bank value of those energy-intensive metals which lose very little weight by corrosion during the useful life of the product of which they are a part. In the case of steel, it is usually a wasting asset during the life of the product, while plastics are often a total write-off on reaching the product.

THE FUTURE

The wide spread of ore reserves, smelter, and semi-fabrication location provides a strength in diversity and a better chance for solutions to be found before the industry declines as a result of failing to adapt. However, I do not believe aluminium will continue to make as much growth by substitution as before, as end-product manufacturers will be wary of becoming too dependent on a single material. At any one time, I suggest, it will be a question of emphasis only and the objective of end-product manufacturers will be to keep products available in several different materials, all of them capable of providing essentially the same characteristics in service. This will tend to favour steady low-growth consumption of raw materials and will make the planning and timing of new smelters more critical.

The effect of environmental impact studies on plant design and location will diminish from the exalted position it holds at present. I consider it will settle down to being part of a normal design and consultative procedure rather than the exotic exercise it is today. A recent study and report on an alumina plant that was approved not only delayed the project by two years but required a report as thick as three London telephone directories distributed to a wide audience. The timber cut down to produce the paper for this report, its drafts, and photocopies was probably more than would be conserved or restored in total on the project over the next 50 years.

I see no pause in the march of government investment and nationalization in the industry. As the companies show more caution over future investment, so governments feel impelled to get things going by intervention. Even those new projects which are funded privately become a target for renegotiation when successful but a duty to be continued when otherwise. All the exposure is downside and the rewards of success are modest. Many of the problems over unbalanced investment could have been solved by joint industry action, maybe by cartels. Yet at the same time that governments act through cartels to enforce their interests, companies are increasingly under scrutiny, both in the USA and the EEC, for the slightest whiff of collusion, constructive or otherwise. This process will continue and will drive more decision-making into the government area.

In that area we also have price and wage control in many EEC countries and currently in the USA. This process will continue and will become part of the growing intermarriage of big industry and government. While it is the received attitude among businessmen to bemoan this tendency, the capitalist-roaders amongst us should at least recognize that the UK aluminium industry has been surprisingly profitable during a long period of price control. It is much easier to tell your customers that the proposed price has been sanctioned by the Government than it is to persuade them of its merit in a competitive climate. Perhaps the process that took place here in the UK on 3 May will reverse this trend for a while, but I think that it is difficult for governments to resist the temptation to 'intervene decisively', and therefore they will continue to do so. This brings us to the LME. Most of the delegates at this conference are familiar with it: you can always detect someone involved with supply of an LME metal by his badly bitten fingernails, a tendency always to locate the nearest telephone before being able to settle down to the business at hand, and a proneness to being absent between 12 noon and 3 pm while other people can only confess to having been out for an Englishman's lunch. It is claimed that the LME has the potential for producing a so-called 'market price' which is beyond the capacity of governments to control. Of course this presumes that consumers and producers believe the same thing and that they like this 'market price'. I find very few who do, except of course those involved in making money on the turn. In the case of aluminium, I think that the quotation will fade from the almost invisible position it has achieved in the last nine months. Any hedge necessary for the supply of aluminium (and I think that the need is modest and will stay so) will be provided by those very same elements of government who are becoming increasingly involved in our industry. I cannot see these new proprietors handing over pricing to be determined around a hole in the floor located in London.

As for the companies themselves: fenced in on all sides, increasingly coralled into narrower and narrower pastures, and cut off from traditional watering holes as they are, I think they will turn to farming — and tenant farming at that. They will no longer take the big risks only to be fenced in again. They will concentrate on their assets in people and systems that can be lent or contracted to the new proprietors, maybe in exchange for metal.

They will act more like service companies to governments and joint ventures, their strength being the ability to evaluate, design, build, and run big, complex, interrelated industries in hostile environments with professionals brought from all over the world, so realizing the dreams of governments and others. They may not do it at a profit, and certainly the discipline of the P & L will not be the manager's first concern when his company is earning a fee. There may be a few wandering cowboys left who will find an unfenced area for a while, but they will probably end up as prospectors for the multinational manager corporations simply because their promises will not be credible in the climate of the future.

The principal sufferer in the year 2000 will be the consumer. He will sink low on the list of the companies' priorities since growth, price, and substitution will increasingly be less important to the manager and warrant less of his attention. The consumer's ability to express his dissatisfaction with one product or another will be circumscribed by the more planned and ordered, but less competitive route by which the product finds its way to him. He will no longer have a champion in the corporation looking for profit from growth and innovation. Maybe he will have to set up an organization to represent him, but even if this occurs (and there is no sign yet of an effective one) it will be more of an input to the existing planning process than an effective demand for diversity and choice (which will be seen as wasteful). Freedom of choice may come to be freedom to punch a computer card — who knows?

In the midst of this predicted gloom there shines a ray of hope for those who did all that research into aluminium buttons and cloth back in the 1950s. We are warned about the dire consequences of too much carbon dioxide being released into the atmosphere by industry—often called the 'glasshouse effect'. Think how much reflective aluminium cloth and buttons would then be needed to survive. Let us hope that if that happens the governments and environmental protection agencies will have had the foresight to create sufficient primary aluminium capacity to satisfy the demand.

M. H. Davies
Copper

The stated objective of this conference is the 'development of a strategy for metals use during the next twenty years'. But for whom is the strategy to be developed? And are all the interested parties represented here?

It is extremely unlikely — even impossible — that a single strategy can be developed which will satisfy the disparate industrial, political, economic, and environmental considerations for any metal, let alone for all, and particularly for copper with which this paper is concerned. The differences in the needs of industrialized nations where the metals are consumed and the developing countries where they are found and mined are very marked. What for example is the correct strategy for importing countries, especially those such as the UK with balance of payments problems? Our own Engineering Materials Requirements

The author looks at the possibilities of developing and expanding the copper market; discusses the present and future availability of copper; comments on its advantages over future competitors; and suggests where its chief markets will lie in the future and to what new uses it will be put.

M. H. Davies is managing director of the Copper Development Association, UK.

Board lists import substitution as one of the main criteria for support of development projects. Again, what is the correct strategy for major industrial fabricators and users concerned about stability of price and reliability of supply? Or for mining companies and state-owned mining industries who want to maximize profits or foreign exchange earnings? How is the right balance to be struck for them in exploiting mineral resources? Should it be done at a high rate, with a low unit cost and a short life, or at relatively lower rate, with a higher unit cost and longer life?

I do not intend to examine these or other alternative strategies for copper, but will concentrate on the strategy that I am required to adopt as director of the Copper Development Association. Our policy is, quite simply, to help develop and expand the markets for copper. It is a policy which can easily be justified on two main grounds. The first is the statistical evidence, i.e. the magnitude of copper reserves and their rate of use—particularly if one is looking at a forward period of only twenty years or so. During and beyond that time technological and economic factors will progressively affect the status of copper as an industrial material, and it is of course necessary that those with a major commercial commitment to copper should continuously review the probable effect of the changes on their interests.

Table 1   Copper production and consumption (all forms) 1976

	Mine production, 10^3 tonne	Total copper consumption, 10^3 tonne	Population, 10^6	Per-capita consumption, kg
Europe	309	3515	368	9·5
Africa	1462	104	384	0·3
Asia	492	1634	1082	1·5
(of which Japan)	(82)	(1409)	108	13·0
USA		2650		
USA and Canada	2200		238	11·7
Canada		235		
Central and S. America	1300	439	318	1·3
(of which Brazil)		(200)		
Australasia	390	152	21	7·2
(of which Australia)	(214)	(150)		
USSR	1200	1250	360	3·5
Middle East	52	26	350	0·07
totals	7698	8815	3229	2·7

The second reason for a valid policy of market development is that the proper medium- or long-term management of copper (or any other resource) dictates that its use should be subject to the twin considerations of fitness for purpose and lifetime cost-effectiveness. Assessment of the latter does of course call for the judgment of Solomon, since it depends on the effects of technological advance and on market forces which influence prices, as well as on money rates which affect the relative burdens of capital and maintenance costs.

The interplay of all the factors involved means that over a period of time materials will find their own level of use and that level will be influenced by the strength of market development and promotional work. Shortages and surpluses will inevitably occur in a free market, owing to changes in the level of economic activity. Such shortages or surpluses will in turn affect prices and therefore material selection policies, and hence the slope of the long-term demand curve. In the mid-1950s, for example, both factors (i.e. supply and price) prompted the partial switch from copper to aluminium for conductors in power and telephone cables. This has progressively reduced actual and potential demand for copper quite significantly.

On the supply side, low prices discourage exploration and the development of new mines and refining capacity, as has happened in very recent years. We will probably all pay for that later on.

In general terms, the commercial development of a mineral resource will depend on the grade, complexity, and location of the ore body — in other words, on the cost of recovery relative to the prospective free market price of the refined metal. One thing is certain: that technological advance or prices which relate unfavourably to potential substitute materials will inevitably lead to substitution. Because of the inventiveness of man, society can and will adapt itself to accommodate any change in the availability of materials and will discover or develop alternative materials or systems to replace those that are in short supply and consequently becoming too expensive. Germany and Britain during the last war provide good examples. Le Chatelier still lives — and his principle of mobile equilibrium is fully applicable.

THE CASE FOR COPPER

The facts which support the case for developing and expanding the usage of copper can be presented as a series of tabulations and graphs.

Table 1 shows:

(a) that mines are concentrated in under-developed countries (excluding the USA,

Canada, and Australia) and that consumption is obviously concentrated in industrialized countries;

(b) a substantial difference in per-capita consumption in relation to state of development — the apparent consumption in Japan (and also in Germany) is high, as a result of exports in the form of copper in finished goods such as cars, domestic appliances, etc.;

(c) the potential, now being realized, for rapid growth in the wealthy Middle East with its very low figure for per-capita consumption (here, too, finished capital and consumer goods containing copper are imported from the West and not reflected in the figure); and

(d) that the USA is broadly in balance, as it was for oil at one time — and that should strike a cautionary note.

In Table 2 we see that the production of refined copper, as distinct from mined copper, has shifted in part to the Western nations, i.e. we import and process concentrates or blister copper. Surely the developing countries will want to add value to their mineral resources and to increase their capacity to produce not only refined copper but also perhaps tube, sheet, and rod copper, and later finished goods. This has happened with other commodities. The participation of Zambia and Chile in contirod production in France and Germany is a step in this direction.

Table 3 indicates the apparent growth in reserves of copper during the last 20 years. It is a warning to those who, misreading statistics, are prophets of doom and gloom. Nearly all forecasts of reserves have been wrong or wrongly interpreted, beginning with the dire predictions of the Paley Commission in 1952.

These figures are *proved* reserves — their augmentation over the years being due to increased exploration and therefore discovery of new ore bodies — and they do not include ocean nodules which, in first-generation sites alone, add at least 200 million tonnes of copper to the reserves. Finally, the resources will of course also enlarge as a result of current exploration; in another twenty years we might well see a higher figure than the present one for total reserves of copper. Although there is naturally a finite limit to reserves, we are a very long way from knowing what it is.

Table 2 Consumption and production of refined copper

	Consumption, 10^3 tonne	Production, 10^3 tonne
Africa	88	1010
Asia and Oceania	1530	1410
Eastern Europe	1610	1730
N. and Central America	2180	2480
S. America	235	572
W. Europe	2548	1400
totals	8220	8590

Table 3 World copper reserve estimates since 1960

Year	Source of estimate	Reserve, 10^6 tonne
1960	US Bureau of Mines	154
1970	US Bureau of Mines	279
1975	US Bureau of Mines	408
1976	Krauss	456
1977	UN study	451

Table 4 World copper reserves by category and average grade

	Copper content, 10^3 tonne	Average grade, %
N. and Central America	143 000	0·68
S. America	113 200	0·99
Africa	76 800	2·80
Asia	47 200	0·75
Europe	8 000	0·67
Centrally planned economy countries	63 000	1·60
World total	451 200	0·96

Table 5   Growth rates of refined copper consumption in
the post-war period (% per annum)

	1953–73	1958–73	1963–68	1968–73
World total	4·3	4·3	3·7	3·5
Centrally planned economy countries	6·3	6·0	5·6	5·9
World excluding c.p.e. countries	3·9	3·9	3·2	2·9

(I have made no mention of scrap which provides about 35% of the UK annual requirements for copper.)

Table 4 complements Table 1 by showing the concentration of reserves in developing countries — a matter for strategic consideration on political and economic grounds. Increased exploration in Europe encouraged by the EEC will surely unearth some workable deposits. The emergence of the Yugoslavian and Polish copper industries is a hopeful sign.

Actual rates of growth in consumption are shown in Table 5. The rates currently being predicted range from 2½ to 3½% per annum for the Western world.

Figure 1, which integrates much of the previous data, is crucial. It shows the lifetime of reserves at different rates of increase in copper production (or consumption). At current rates of production (or consumption) we have a predicted life in excess of 50 years even if no more copper is discovered — and that excludes ocean nodules, which extend the static life index of reserves to about 80 years at current rates of consumption. At annual growth rates of consumption of 3%,

for example, there are still about 50 years of proved reserves, but clearly more copper will be found.

Figure 2 gives reserves as a function of price, at 1975 figures; today the curve should be displaced upwards because of the increase in mining costs due to inflation and other factors. As the selling price of the metal increases, however, so it becomes profitable to work lower-grade deposits.

Of brief interest are the actual movements of average price in the USA and on the London Metal Exchange owing to market forces, shown in Fig. 3. Although there are wider fluctuations in the LME price than the US producer price, the broad trends are similar.

On the basis of that factual evidence, is it realistic to talk about a strategy for copper, couched presumably in terms of conservation, when there is enough (in the absence of catastrophic and unforeseeable change) to last at least fifty years and probably a great deal longer? So if I have laid a satisfactory base for the market development strategy of the Copper Development Association on

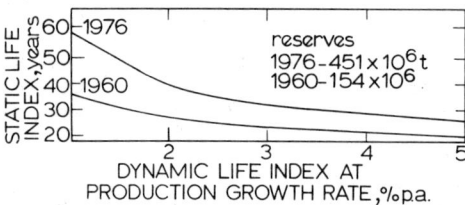

1   Lifetime of copper reserves as estimated 1960 and 1976: static vs. dynamic life indices

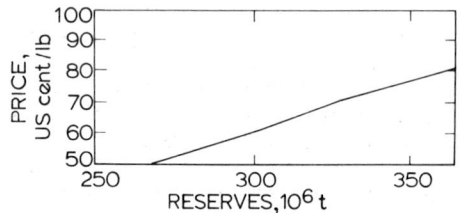

2   World copper reserves as function of price (estimated 1975)

Table 6   Per-capita consumption of copper in selected
industrialized countries

	1963	Average 1972–74	Population, 10^6
Fed. Republic Germany	8·90	11·47	62
Italy	4·50	5·38	55
United Kingdom	10·40	9·34	56
Japan	3·67	9·34	104
United States	8·40	9·82	203
USSR	3·27	4·31*	133
Asia (excluding China)		1·50	1082
Mid-East (incl. Pakistan)		0·07*	350

*1976

behalf of the producers I would like to spend a few minutes expanding on that strategy. The questions to be asked are: where and what are the markets for copper and how will they grow? What are the opportunities and what are the challenges from substitute materials or systems?

CHALLENGES AND OPPORTUNITIES

Let me first deal briefly with some of the challenges, recognizing that patterns of use will inevitably change as a consequence of technological and economic forces and that real advances in technology will produce growth or decline in use for copper in existing and potential new applications. Some of the past, present, or future threats to copper include: aluminium for electrical conductors; fibre optics for telecommunication systems;

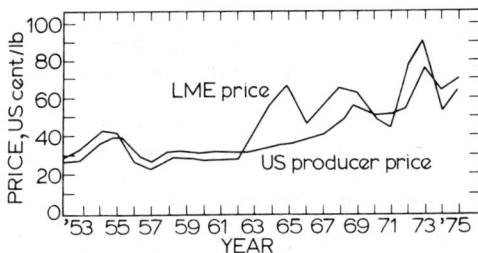

3   **LME and US producer prices for refined copper**

plastics, including metallized plastics, for undemanding space-filling applications and, increasingly, for engineering use as they become more reliable; titanium for heat exchangers, desalination and chemical plant, and miniaturization and micro-miniaturization; and plastics and aluminium for car radiators. With respect to two of these contenders for copper's markets, however, it must be remembered that aluminium is very energy-intensive and that plastics are derivatives of oil or, even more expensively, of coal.

Now what about the opportunities? Firstly, a trend towards higher-quality, longer-lasting materials must favour copper. What we need to remember are the technical virtues of copper and its alloys:

(a) excellent heat and electrical conductivity
(b) excellent corrosion resistance
(c) excellent fabrication, machining, and joining properties etc.
(d) non-fouling properties in sea water
(e) a very wide range of available shapes, especially extruded copper and brass sections.

These are the characteristics which will determine its future use, together with its price relative to competitive materials. In general, as standards of living rise throughout the world and particularly in the populous developing countries, so will the consumption of copper increase.

Table 6, showing the variation in per-capita consumption, illustrates the capacity for normal growth not only in Asia, the Middle East, and Africa as their standards of living

Table 7   Copper consumption by end-use sector
(% per annum)

	USA 1974	W. Europe 1975	Japan 1975
Electrical and electronic products	46·3	54·3	52·0
Building construction	15·9	15·5	8·8
Transportation	10·1	10·7	17·1
Industrial equipment	18·8	14·0	15·0
Consumer products	8·9	5·5	7·1
totals	100·0	100·0	100·0

rise, but also in Europe. Contrast the figure for Italy (or Spain, Yugoslavia, etc.) with that of the UK and Germany. If India, with a population of 600 million, were to increase its present per-capita consumption of about 0·13 kg to Western levels, world demand would rise by another 5 million tonnes per annum.

The importance of Europe is shown in Fig. 4. In the USA consumption of refined copper has been virtually static over the last 12 – 14 years. Europe, on the other hand, is a larger and still growing market; it must become the focus of attention of the producers, especially those in Africa and South America. In the USA and the UK, however, imports (and exports) of finished goods containing copper mask the true figure for per-capita consumption.

The existing pattern of consumption by end-use industries can be seen in Table 7. There are no major differences between the regions, but the Japanese, for example, have a somewhat lower rate of use in building because of local custom and tradition.

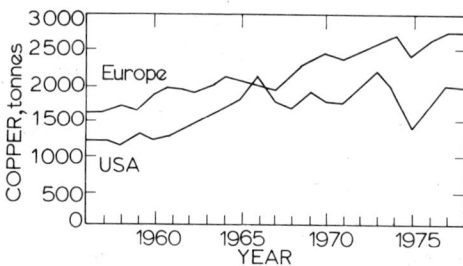

4   Consumption of refined copper 1957–77

Looking ahead at some of the prospective markets which will take advantage of the properties of copper there is, for instance, solar heating. The mass market for solar heating systems will of course be determined by the balance of costs and benefits relative to the costs and availability of fossil fuels. The systems which are being installed in the UK are predominantly copper.

Cupronickel for ships' hulls will improve fuel economy and reduce maintenance costs. The first UK-built cupronickel-hulled vessel was launched in 1978. Two more are being built. These vessels have solid hulls of $^3/_{16}$ in plate, but large vessels will of course have clad plate.

Shape-memory-effect alloys offer exciting prospects for brass.

Fish farming is another interesting growth industry, for which again it is the anti-fouling properties of cupronickel which are attractive. While there is 'apparent' fouling even of cupronickel cages, the shell and weed do not attach to the cage itself but to debris mechanically trapped by the mesh and can readily be cleaned off. Other important marine applications include duct linings and desalination plant.

One very large potential market to which I should refer is the electric car, which might use five times as much copper as its petrol-driven counterpart, e.g. 100 kg compared with 20 kg per vehicle. However, the advent of the electric car on a massive scale is scarcely dependent on improvements in copper technology, but rather on development of a lightweight, cheap battery and on environmental decisions such as the banning of petrol-driven vehicles from city streets.

The scope for copper as an essential trace metal in food and animal production, as well as the control of disease, is also promising, and there are many other interesting possibilities based on the technical merits of copper vis-à-vis its competitors. In addition, the copper metals have great beauty and aesthetic appeal. The mosque in Regents Park is a very impressive example of the use of 'Incracoat' for roofing. This is a laminate of copper alloy and polyvinyl fluoride intended to preserve its natural colouring for many years.

In conclusion, let me repeat that copper is not scarce now in resource terms, although there may be temporary shortages in supply. It is not likely to be scarce for many years, and twenty years hence it is more than likely that the proved reserves will be sufficient for more than 50 years of estimated consumption. Western Europe is the cockpit. Even though the potential for growth is larger in countries such as India, the pace of economic growth and consumer demand is slow, whereas Western Europe is an established large and growing market. More importantly it is (along with the USA and Japan) the area where the main battles for markets, based on the application of present technology, will be won and lost. The forces of the copper industry have not yet been properly marshalled for those battles. But they will be!

SOURCES
The main sources of my data are the Metallgesellschaft, the World Bureau of Metal Statistics, the Commodities Research Unit, and a UN study 'Copper: the next fifteen years', published in 1979 by the D. Reidel Publishing Company. Although some of the data are a little old, their dating does not weaken their support of the argument.

D. A. Robins
Tin

If the importance of a metal were judged solely by the tonnage produced or consumed, then tin would rank as a minor metal, since in 1978 the world production (excluding the People's Republic of China and the USSR) was only approximately 195 000 tonnes. In fact, however, tin has an importance quite unrelated to the relatively small size of the industry for two reasons.

Firstly, tin is a strategic metal with a vital part to play in many important industries. Whereas there is a closely defined tin mining industry concentrated in certain relatively small areas of the world, there is no clearly defined tin-using industry. Rather, many different industries make use of tin where it constitutes often quite a minor part of the product or process, although the role of the tin is often of vital importance. The strategic nature of tin has meant that a country such as the USA with no indigenous source of tin has maintained over the years a strategic stockpile of the metal.

Secondly, tin has outstanding international and political significance. The metal is mined almost exclusively in developing countries, some of which have economies highly dependent on the export earnings from tin. On the other hand, the developed countries who consume the bulk of the world's tin have little or no domestic production. With no other metal is there such a clear division and such interdependence between the developed and the developing worlds.

Although tin is a comparatively low tonnage metal, it has a vital part to play in current technology and leads the world in international co-operation on commodities. At current rates of consumption tin should be readily available for at least the next 50–150 years and its major uses are likely to be retained, despite attempts at substitution. Improvements in technology such as the development of the two-piece tinplate can are likely to ensure that tinplate continues as the premium packaging material for food.

The use of tin in solders for the electronics industry is particularly cost-effective, and new uses likely to expand include the development of organotin compounds. With the working of progressively leaner ores and improvements in living standards in the producing countries, it is envisaged that the price of tin in real terms should progressively increase.

The author is Director of the International Tin Research Institute, Greenford, Middlesex.

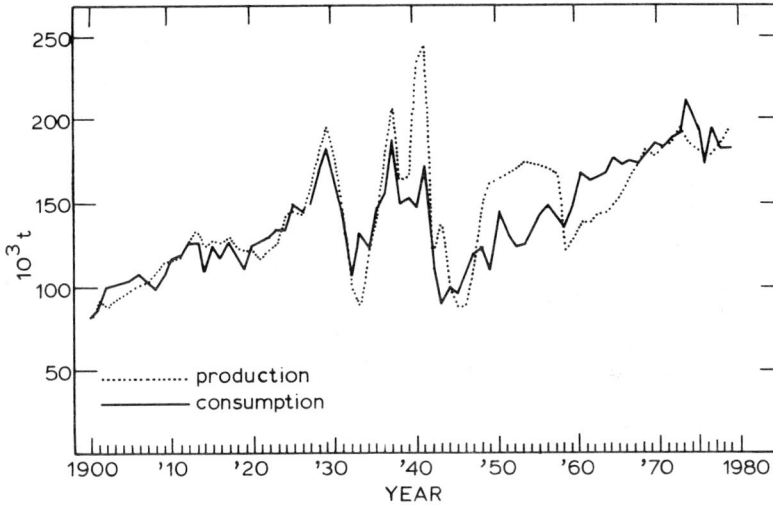

1 World tin production and consumption since 1900 (excluding People's Republic of China and USSR)

Any attempt to develop a future strategy for the exploitation of the world's metal resources must, when applied to tin, take into account both the vital part tin plays in many different industries and the international implications of such a strategy.

AVAILABILITY

The annual rate of growth of the tin industry over the years has been relatively low in comparison with most of the main tonnage metals (Fig. 1). In 1978 world production and consumption was only about 10% higher than it was 50 years earlier. It is therefore reasonable to expect the future growth rate of the industry to be also relatively slow. This is important when attempting to assess whether or not tin will still be readily available in the next century. The relatively low growth rate of the tin-producing industry is somewhat misleading, however, since it conceals major changes which have taken place over the years, and in particular the mining of progressively leaner tin ores. This is true for both the underground mining of vein deposits in Bolivia, for example, or the treatment of alluvial deposits in an area such as Malaysia by dredging or gravel pump mining. Most

of the cost of tin production arises from the working of ore bodies which, in the case of alluvial mining, may contain as little as 0·01% tin as cassiterite.[1] This trend towards the exploitation of lower-grade deposits will continue in the future, as also will the development of economic mining at greater depths and the growth in importance of offshore mining.

Many attempts have been made to estimate the world's resources of tin ore, but according to Professor Hosking[2] ' . . . neither tin producers nor tin users should lay much store by the various published estimates of the world's tin reserves and resources, as they may well be very wide of the mark'. The assessment of possible future availability is made more difficult by the normal practice of mining companies in the developing countries of usually seeking only to confirm the presence of workable reserves for a period of 15 or 20 years ahead. The remoteness and lack of surveys of some of the areas involved also make estimation difficult and it is impossible to quantify with any accuracy the future availability from offshore deposits. Also, the decision whether or not a particular ore deposit is economically workable depends on

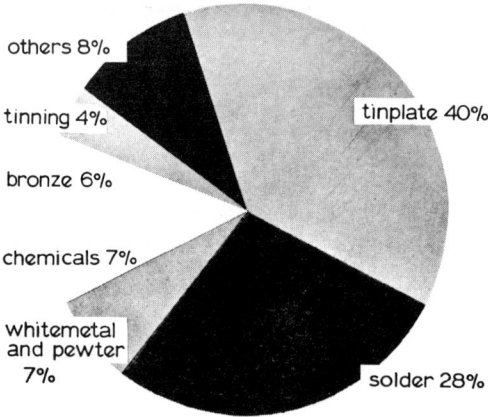

2 **Uses of primary tin**

others 8%
tinning 4%
bronze 6%
chemicals 7%
whitemetal and pewter 7%
tinplate 40%
solder 28%

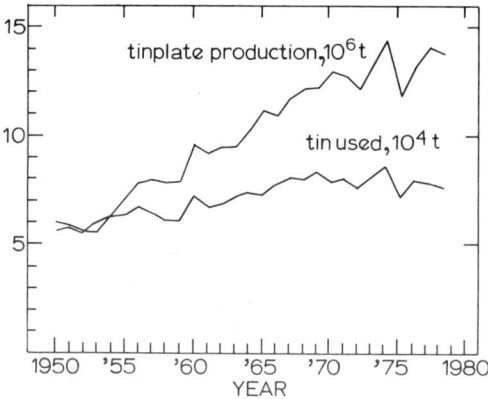

3 **World tinplate production and tin used**

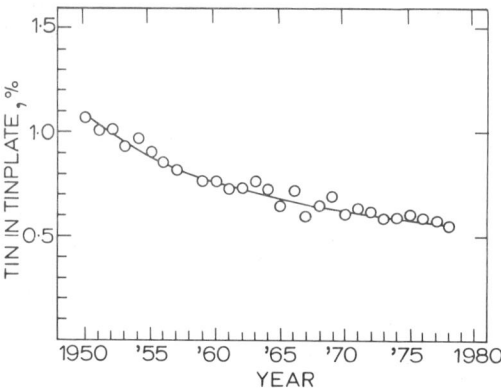

4 **Average tin content of tinplate**

the tin price, so that waste ground at one price level can become an ore body at a higher price level. A comprehensive analysis of world tin resources attempted a few years ago estimated that, assuming current consumption trends, tin would be readily available for at least 50 – 150 years.[2]

CONSUMPTION TRENDS

Tin's unique combination of properties makes it suitable for many different applications. In each specific case, however, the technical advantages gained by its use must justify the cost involved.

In the last few years the price of tin expressed in sterling has increased significantly although how much of this increase is due to inflation and the decreasing international value of the pound is a matter for debate. To an organization like the International Tin Research Council which undertakes research and development work to help maintain a profitable market for tin, it is encouraging to see that in 1978 world tin consumption was slightly higher than in the previous year despite these higher price levels. This resilience of the tin-using industry stems from the continuing development of new and improved uses for the metal, some of which are highly cost-effective.[3] In this paper it is possible to consider briefly only one or two examples of main areas of tin usage as a guide to possible consumption trends.

Although the range of uses of tin has widened in recent years, the largest single use continues to be as a coating on steel strip to form tinplate, accounting for about 40% of world consumption (Fig. 2). Any future strategy for tin must therefore take into account the possible trends in the tinplate industry. The use of tin in solders accounts for about 28% of the total, with many other different uses making up the remainder. This overall pattern of consumption is only likely to change slowly, with the percentage of total use formed by soldering and other uses possibly increasing at the expense of tinplate.

Tinplate

The total world production of tinplate has increased over the years and is likely to continue to increase in the future as the world population increases and living standards improve (Fig. 3). Owing to changing techno-

Table 1   Energy consumptions

	Energy consumed in producing raw material for 1 tonne, GJ	Number of containers per tonne	Energy consumed per container, kJ
Tinplate can	49	16 500	3 010
Aluminium can	395	44 500	8 660
Bimetallic can	77	18 400	4 210
Glass bottle, returnable	54·8	2 000	27 540
Glass bottle, non-returnable	54·8	4 000	13 770

logy, however, the requirement for tin has not increased at the same rate as tinplate production. The percentage by weight of tin used in tinplate in the early 1950s was about 1% and this has progressively decreased to half this value (Fig. 4). This more efficient use of tin, however, has enabled tinplate to compete successfully with possible alternative packaging materials. Some further decrease in the world average coating weight is to be expected as the latest technology becomes more widespread, but from a technical point of view there could well be a lower limit to this trend.

A feature of the tinplate industry has been the changing pattern of world production. Whereas in 1950 only five countries produced more than 100 000 tons a year, there are now 19 countries doing this, and this number will further increase in the future. The bulk of the world's tinplate is used for the preservation of food in cans and, with increasing energy costs, canning has significant advantages over possible alternative procedures such as freezing food and keeping it frozen until consumption. Table 1 compares the energy consumed in making containers in different materials, and the merit of tinplate is immediately apparent.

Overall economies in the amount of steel required to make a given number of cans are likely to be gained from the more widespread use of the new two-piece drawn-and-ironed can. In manufacturing this type of can the steel is reduced in thickness and is therefore in the worked condition so that thinner material can be used. The tin coating still has an important part to play in that, in addition to its other advantages, it acts as a solid lubricant during the wall-ironing process. Although at the moment confined mainly to beverage cans, this new type of container is likely to be introduced gradually for food products. The presence of the tin coating is therefore liable to lead to a cheaper product because of the reduction in the amount of steel required.

Current trends coupled with recent technical advances confirm the view that tinplate will remain an important packaging material well into the next century.

Soldering

When brought in contact with many clean metals, molten tin reacts to form intermetallic compounds at the interface; when making use of this property in a soldering operation, the tin is normally diluted with lead, which is inert. When using tin–lead solders there is always an economic incentive to use the highest lead content possible and to reduce the size of the actual joint. With the higher tin price levels in recent years some of the older inefficient uses of solder have been lost to alternative procedures, but this loss has been offset by the increasing use of solder in the electronics industry. The large number of joints on a printed circuit board can be rapidly made by passing it through a wave of molten solder. Provided the surface conditions of the board itself and of the various terminations are properly controlled, hundreds of reliable joints can be made in only 2 or 3 seconds. The composition of the solder is required to be near the tin–lead eutectic, and in practice 60% tin and 40% lead is normally used. As many manufacturers have found by

65

hard experience, attempts to economize by reducing the tin content can be very costly in terms of faulty products.

This use of tin in the electronics industry is highly cost-effective and despite miniaturization, which makes the use of tin even more worthwhile, the amount of tin used worldwide is likely to increase. How far this will be offset by the continuing loss of some of the less effective solder uses is an open question.

Other uses

The wide range of different tin uses making up the remaining one-third of consumption include those with a very long history, such as tin-bronze, alongside relatively new applications such as the addition of tin to cast iron to give a fully pearlitic structure. Some of the newer uses involve a relatively low percentage of tin and the future trend is likely to be towards this type of application. An alternative approach is the design of processes in which a large quantity of tin may be involved but is not actually consumed. An example of this is in the manufacture of float glass using a large bath of molten tin in an inert atmosphere.

The most rapidly growing use for tin at the present time is in chemicals, and particularly organotin compounds. Figure 5 shows the rising consumption of organotin compounds in the USA in recent years. It is interesting to note that a PVC bottle stabilized with an organotin compound may contain about the same percentage by weight of tin metal as a tinplate can. In the field of agriculture, the use of organotin compounds has important environmental advantages and is an expanding area for tin consumption.

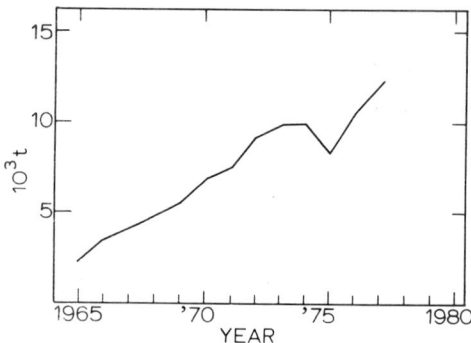

5 **Consumption of organotin compounds in USA**

RECYCLING

The cost of tin to the user has always encouraged reclamation and recycling wherever possible. In the bronze industry, for example, something like half of the production is from secondary metal. There is also considerable use of secondary tin in the solder industry, and in the tinplate industry scrap arising during can-making is always de-tinned. The tin on used cans, however, is for the most part not reclaimed, nor is much of the tin used in tin chemicals.

The overall extent of recycling and reuse is not easy to assess, for lack of reliable statistics and also because of the tendency to concentrate these statistics on the reclamation of tin metal rather than the reuse of the tin content of tin-containing materials. The published data for the USA are probably the most comprehensive: in 1977 approximately 78% of consumption was primary tin while 22% was recycled metal.

It might at first be assumed that higher tin price levels would encourage increased recycling, making marginally uneconomic processes economic. Except perhaps in the very short term, however, this is not likely to be so. The continuing use of tin-containing materials at these higher price levels means that the amount of tin per unit decreases and the complexity of the application increases. This makes reclamation and recycling more difficult. An example is the de-tinning of used cans when reduced coating thicknesses have been introduced together with improved and more adherent lacquer coatings on top of the tin. Attempts made in many countries to de-tin used cans extracted from domestic refuse, with the object of reclaiming both tin metal and steel free from tin, have met with little commercial success.

An alternative approach not involving a separation of the two components could well prove more viable. No attempt is made to de-tin copper wire or copper sheet since the tin content is best reused by melting down the composite material to form bronze. In a similar way some of the tin-bearing ferrous scrap is already used in the cast iron foundry where the presence of tin is beneficial in certain applications. A possible future outlet for this scrap is in the manufacture of sintered iron parts. Whereas a high tin content is not acceptable in cast and wrought steel, it is of benefit in certain sintered iron parts, and additions of 1, 1·5, or 2% tin are used commercially to give better dimensional control and/or lower sintering temperatures. It has already been demonstrated experimentally that iron powder made by melting and atomizing used cans is acceptable for this type of application, additional tin being added to make up the required composition. It is appreciated, of course, that so far the tonnages involved in powder metallurgy are not sufficient to absorb all the tinplate scrap available.

As the various uses for tin become more complex, so the economics of recycling becomes less attractive. The use of tin-rich white-metal alloys is one field where recycling was and still is practicable for large marine bearings. Now, however, many bearing applications, although still making use of the good antifrictional properties of tin, involve complex or composite materials where recycling is not possible. It seems likely that in the foreseeable future the contribution to overall tin consumption from recycling will not increase significantly and indeed might well decrease as the range of uses and complexity increases.

FUTURE STRATEGY

Serious attempts are being made through the UN Conference on Trade and Development to extend international co-operation on commodities and, in particular, to set up a Common Fund, one function of which will be to finance the holding of buffer stocks. At the present time, however, tin is the only metal which is the subject of a successful commodity agreement, possibly because of the clear division between the producing developing countries and the consuming developed countries already referred to. The Fifth International Tin Agreement in force at the present time has been ratified by all the major tin-producing countries of the world and most of the consuming countries including the UK, the USA, and the USSR. This means that there is already an agreed international policy on tin as laid down in the twelve Objectives in Article 1 of the Agreement.[4]

These Objectives give high priority to the needs of the developing world. After the first and second objectives, which are the balancing of the supply and demand and the prevention of excessive price fluctuations, the third objective is 'to make arrangements which will help to increase the export earnings from tin, especially those of the developing producing countries . . .' The significance of this in more positive terms is brought out in objective (d) which calls for conditions 'which will help to achieve a dynamic and rising rate of production of tin on the basis of a remunerative return to producers . . .' and objective (f) 'to improve further the expansion in the use of tin and the indigenous processing of tin, especially in the developing producing countries'. The objective of increasing export earnings can also of course be achieved by selling the present tonnage production at higher price levels.

The indications are that in the longer term the real price of tin will increase. This increase will occur as a result of both the working of progressively leaner ores and the improving of living standards in the producer countries. Ideally, marked fluctuations in the price of tin should be avoided since this makes investment planning difficult, but maximum use of the world's tin resources will be encouraged by a continuing and progressive increase in the real price. To maintain a strong tin market under these conditions will require, from a technical standpoint, the refinement of the established uses and the development of new ones taking maximum advantage of the metal's unique properties. Increasing still further the diversity of uses will also help protect against possible substitution and changing industrial patterns. The future for tin, therefore, is seen as one of slow but steady growth in world consumption, with the more cost-effective applications becoming increasingly important.

REFERENCES

1 'A Guide to Tin', ITRI Publication No. 540
2 K. F. G. HOSKING: 'The search for deposits from which tin can be profitably recovered now and in the foreseeable future', Fourth World Conf. on Tin, I, 25, London International Tin Council
3 D. A. ROBINS: 'Technological developments in tin consumption combat substitution', ITRI Publication No. 541
4 Fifth International Tin Agreement (International Tin Council, 1 Oxendon Street, London SW1)

A. R. L. Chivers
Zinc, cadmium, and lead

This conference comes two weeks after the seminar 'Zinc in Europe' organized by the Zinc Development Association with the support of the European Zinc Institute. That meeting attempted to look ahead only as far as 1985 and was mainly concerned with the European scene although, since zinc is a truly international commodity, data and estimates were presented for the whole of the Western world. This limitation is itself a source of uncertainty. Will the Chinese develop their extensive zinc deposits and, if so, will they refine zinc and use it in their own industrialization programme or sell it abroad? Presumably the same question mark hangs over lead.

In 1978 Western world zinc mine production was 4 719 000 tonnes metal content, 100 000 t below the record figure of 1977. By 1983 this total is expected to have reached nearly 5 300 000 t, an estimate based upon published data for projects already started. Of the new mine capacity 40% will be in developing countries which at present account for 28% of total mine production. Many other new mining projects are under consideration.

There are sufficient zinc ore reserves in Canada, Australia, South Africa, and Europe to last for 13 years if consumption in the Western world grows by 2½% per annum from 1978 levels. Reserves in the Socialist countries are nearly the same. Zinc production capacity continues to grow, despite the fall-off in demand since the boom of 1973-74, so the industry's main concern is in developing its markets; the prospects for growth are briefly considered . . . Cadmium is purely a by-product of zinc and recovery efficiency is being improved. No shortage is expected unless a new major market appears . . . The lead industry makes much use of recycled material; the main expendable use nowadays is for antiknock additives which will gradually be phased out of petrol if present trends continue. Motor vehicle batteries account for 50% of current consumption and this market continues to grow. Western world lead ore reserves will be adequate for more than 25 years if the amount mined increases by 2½% per annum. For both lead and zinc new reserves are expected to be discovered during this time. The rate of consumption of both metals is growing fastest in the developing countries.

The author is with the Zinc and Lead Development Associations, London.

The total of 73 800 000 tonnes of zinc in indicated reserves in the Western world will last for 13 years if consumption in the West increases at 2½% per year from the 1978 level. The figure for reserves is nearly doubled if the Socialist countries are included. As the reserve figure ignores deposits where zinc is a by-product, and further reserves are likely to be found in the next decade, there is no sign of any general shortage of zinc. Of these reserves, 80% are in Canada, Australia, and South Africa and about 5% in Europe. Mining costs differ considerably from one deposit to another, depending on the grade of ore, situation, wage rates, taxation policy, etc. The European mines, which are generally small, are among the dearest to operate.

Those who are sceptical about finding new reserves should note that the Tara mine about 30 miles from Dublin is Europe's largest zinc mine, currently producing about 200 000 t of zinc content a year and about 50 000 t of lead. It was discovered, very near the surface, within the past 15 years.

Since the boom years of 1973–74 there has been an excess of zinc smelter production capacity over consumption, many plants operating at well below their maximum output rate. Between 1980 and 1985 another

half-million tonnes of production capacity is expected to be available bringing the total to 5 600 000 t (6 300 000 t nominal) compared with 4 300 000 t actual production in 1978. This is a growth in productive capacity of 2·1% per annum approximately. If zinc consumption grows at 2·1% p.a. after 1980 the present over-capacity of about 640 000 t will grow to 700 000 t by 1985. Growth in consumption at 3·5% from 1980 would reduce overcapacity to 400 000 t, but this rate of growth is not considered likely in view of current economic forecasts. The zinc industry is more concerned with developing markets than with availability of future supplies. The position might be reversed eventually if there is a shortage of aluminium because of its extended use in car engines.

In round figures, the main uses of zinc are for coating steel (40%) and for brass and zinc die casting (each 20%), while zinc oxide and rolled zinc each take less than 10%. The market for zinc-coated steel strip for cars in North America is expected to grow with the introduction of legislation to delay the onset of rusting. Against this, galvanized strip for building construction may be replaced by zinc–aluminium alloy coated steel using only one quarter the weight of zinc per square

ZINC STATISTICS

	1978	1983	1985
Free world mine production, t Zn	4 700 000	5 300 000	—
Free world metal production, t	4 300 000 (actual)	—	5 600 000 (capacity)
Free world slab zinc consumption, t	4 500 000	—	5 070 000 at 2·1% p.a.
			5 720 000 at 3·5% p.a.

NEW MINES OPERATIONAL 1983

Europe	168 000
Africa	32 000
America	300 000
Asia	58 000
Australia	70 000
total	628 000 t Zn

total　628 000 t Zn　　*Figures from International Lead and Zinc Study Group*

INDICATED ZINC ORE RESERVES

World total	135 000 000 t Zn	*Figures from*
Free world	74 000 000 t Zn	*the Organization for Economic Co-operation and Development*

metre. This trend is already well advanced in Australia, a large market. Brass is mainly used in plumbing fittings, electrical components, and heat exchangers, where its metallurgical properties are needed. Memory alloys may open new markets for thermostats, fasteners, and toys.

Zinc die castings have to compete mainly with plastics, and have lost many largely decorative applications, especially in US cars. They have much scope for replacing machined components now made in steel or brass, and assemblies of steel pressings. New foundry alloys may secure some markets now supplied by malleable iron and bronzes. Rolled zinc is mainly used for battery cans. Cells using zinc powder anodes are growing in popularity; since these use zinc more efficiently, they may retard growth in total zinc consumption for batteries.

The opportunities for recycling zinc more effectively are not great. More will be recovered from steelworks when old galvanized material is melted. The fume can be fed to an Imperial Smelting furnace. Brass is already recovered on a large scale and the grey zinc oxide from brass works is used in chemical manufacture. Zinc die castings are recovered as zinc or zinc dust from fragmentation plants handling scrapped cars and domestic appliances, but as the die castings get thinner the recovery will be less effective. Zinc used in pigments and batteries is lost, eventually to re-enter the earth and surface waters, where it is an essential element for all forms of life.

CADMIUM

Cadmium is virtually produced only as a by-product of zinc (with a little coming from lead smelters) though some is recovered from large nickel–cadmium batteries. Free-world total production of cadmium in 1978 was 12 800 tonnes. Antipollution legislation has the effect of raising even higher the recovery of cadmium in zinc refining, while at the same time tending to restrict the uses of cadmium and its compounds. A shortage of cadmium will only arise if a major new use develops, e.g. for solar cells. The main uses now are in colours and stabilizers for plastics, in batteries, and in plating.

LEAD

The lead industry is far more dependent on recycling than is zinc, and the trend is bound to intensify. The major expendable use of lead is in antiknock additives, which will gradually be phased out of petrol if present trends continue, despite their contribution to the efficiency of high-octane fuel production. Lead pigments for paints are already a very small part of consumption. The widest use of lead nowadays is in SLI batteries for cars and commercial vehicles; it accounts for half the lead market and is growing. No alternative battery is expected before the year 2000, and probably not even by then. The usage of lead for cable sheathing has declined in the past decade, notwithstanding the new plants installed in developing countries whose

LEAD STATISTICS

Free world mine production 1978, t Pb	2 500 000
Free world metal production 1978, t	3 700 000
Free world metal consumption 1978, t	3 700 000

Figures from International Lead and Zinc Study Group

INDICATED LEAD ORE RESERVES

World total	150 000 000 t Pb
Free world	90 000 000 t Pb

Figures from the Organization for Economic Co-operation and Development

products have replaced exports from the industrialized countries. Much old lead sheathing is recovered, together with the copper or aluminium conductors. Lead sheet and pipe are important in only a few countries.

Lead mine production in the Western world in 1978 totalled 2 500 000 tonnes metal content, of which more than half came from North and South America. Lead metal production was 3 700 000 t, which illustrates the importance of scrap recovery. Consumption was in close balance with production.

Total indicated world lead reserves are estimated at 150 000 000 t lead content, of which at least 90 000 000 t are in the free world, enough for over 25 years if mine output rises by 2½% per year. New mine projects are expected to yield nearly 300 000 t of lead in concentrates annually by 1983, but against this must be set some mine closures. Another 250–300 thousand tonnes could be available by the end of the '80s if the demand develops. Lead smelter projects, including secondary lead refineries, will provide additional capacity of about 300 000 t/year by 1983. Another 350 000 t annual capacity is known to be under consideration for possible introduction later in the decade.

The main concern of the lead industry is the maintenance of the SLI battery market.

Presumably the trend towards smaller cars in the USA will mean smaller batteries and further efforts will be made to reduce the lead content of a battery for a given energy storage—much has already been achieved in recent years. Any loss of lead consumption here may be made up by using battery-powered delivery vans more extensively; the UK is far ahead of most other countries in this respect.

CONCLUSION

Any attempt to assess production and consumption of these three metals in the year 2000 can be only a guess. Much will depend on how we have maintained our power supplies and whether we have learned by then to employ the readily accessible resources of the world rather more sensibly than we do now. The rest of the century is likely to see a catching up of standards of living in the developing countries, where demand for lead and zinc has been growing much faster than in the industrialized countries.

P. L. Dancoisne
Manganese

As a result of the work of R. Mushet in 1856, manganese has become an essential element in steelmaking because of its ability to overcome the disadvantages of sulphur content, avoiding hot shortness, and also to deal with deoxidation in steelmaking processes. In addition, manganese confers many properties on ferrous products. Manganese usage is developing to improve tensile strength, hardenability, and hot strength together with the advantages of an austenite-forming element in stainless steels and Hadfield steel.

This brief review of manganese consumption by the steel industry, which accounts for over 90% of the manganese market, presents data concerning the manganese ore market as well as manganese production. Reserves and resources are discussed but difficulties in reaching accurate estimations are pointed out. Reserves are sufficient to meet actual world requirements for the next fifty years assuming a 4% growth rate of manganese consumption. In the future, new deposits are likely to be discovered and within 20 years or so manganese from sea bed nodules could supply part of the world demand.

The author is director of the Manganese Centre, Paris.

Manganese has played a key role in ushering in the modern steel age and will remain an indispensable element for the steel industry. It is an essential addition to nearly every grade of steel. Today, nearly 90% of manganese production is used in the steel industry. Other applications, such as its use as an alloying element for aluminium, copper, dry cell batteries, and chemical compounds, represent a minor sector of the manganese market and are not relevant to this paper.

After briefly outlining manganese consumption, the sources of this metal and their ability to satisfy requirements during the next fifty years will be discussed.

CONSUMPTION

Manganese is predominantly an alloying element used in steel and other metals, in the form of ferroalloys or manganese metal. High-carbon ferromanganese is the commonest ferroalloy, representing 80% of consumption. A typical analysis is 73-78%Mn, 12-15%Fe, 7·5%C. Low-carbon ferromanganeses are also available and are essential alloying elements in certain types of steel. Silicomanganese with 65-68%Mn, 16-21%Si, 1·5-2%C, balances Fe, purer qualities of manganese, massive manganese, electrothermic manganese, and electrolytic manganese are other commercially

Table 1   Breakdown of Mn production by type of alloy (in order of magnitude)

Type of alloy	Production, 10^6 tonne
High-carbon Fe–Mn	3·5
Si–Mn	1·0
Low-C and medium-C Fe–Mn	0·4
Mn metal	0·06

available types. The comparative world production of these alloys is given in Table 1.

A simple means of estimating the steel industry's manganese requirements is to consider the specific consumption per tonne of raw steel produced, i.e. the quantity of manganese introduced in the form of ferro-alloy to each individual tonne of crude steel. This consumption depends on numerous factors, the most important of which are the steel grade and the steelmaking method. In industrialized countries the current specific consumption of manganese amounts to 6·2 kg per tonne of steel. This may be divided up roughly as follows: 4·3 kg of manganese for high-carbon ferromanganese, 0·9 kg for silicomanganese, and 1 kg for low-carbon qualities and manganese metal. This breakdown differs from country to country, as for example in Japan where the proportion of silicomanganese is higher. These figures are averages; for alloyed steels the consumption is nearer 10 kg of manganese per tonne of raw steel, approximately 6·5 kg being for high-carbon grades and 3·5 kg for silicomanganese and low-carbon grades.

Consumption varies widely with the steel-making process employed and the development of the basic oxygen process in place of the open-hearth process has, since the 'sixties, led to a quite considerable drop in the consumption of high-carbon ferromanganese. In Europe and in the United States, for example, it fell from 5·3 kg per tonne of steel in 1967 to 4·3 kg in 1977. Consumption also fluctuates with the economic situation, since this influences the type of steel grades produced. Nowadays the effect of the development of the basic oxygen process is no longer a factor in the level of consumption; apart from fluctuations due to economic causes, one may say that this has stabilized.

During recent years the understanding of phase transformation in steel has led to the possibility of combining high strength and high fracture toughness to produce materials of enormous potential for making lightweight structures of enhanced efficiency. These HSLA and dual-phase steels contain over 1%Mn and will no doubt lead to a slight increase in manganese consumption in the future.

For the purposes of this study, it seems fair to assume that manganese requirements will increase at a rate at least equal to that of steel production, bearing in mind that when various alloys used in the steel industry are manufactured from manganese ore, a certain quantity of manganese is lost in the slag, etc. Furthermore, it is worth noting that manganese is used in the form of ore to increase the manganese level of pig iron in the blast furnace in cases where the manganese content in the iron ore itself is not sufficient. To this end, low-grade and generally ferruginous manganese ores are used.

ORE MARKET

World consumption of manganese ore (all categories) reached 25 million tonnes in 1977 compared to 18 million tonnes in 1970, approximately 10% representing low-grade ores used directly by the steel industry.

Manganese is a relatively plentiful element of the earth's crust, occupying twelfth position in the order of abundance with an average content of 0·1%. However, the number of manganese deposits which constitute sources of commercial ore are relatively few and consequently there are few producers. There exist at present five main producers of manganese ore to satisfy the needs of the free world. Table 2 shows the

Table 2   Manganese ore production in years 1970, '75, '77 (10^3 tonnes)

Producer country	1970	1975	1977
S. Africa	2 679	5 897	5 831
Gabon	1 615	2 118	2 086
Brazil	1 878	2 115	1 904
Australia	750	1 555	1 995
India	1 820	1 360	1 633
USSR	6 841	8 459	8 595
China (estimated)	900	1·4	1·6

Table 3 Manganese ore imports (all grades) in
1970, '75, '77 (10^3 tonnes)

Importing country	1970	1975	1977
USA	1 711	1 459	848
Japan*	2 584	3 697	1 544
France	1 049	1 157	928
Germany*	709	654	455
Norway	586	1 083	590
UK	524	253	328

*Including ferruginous ores

outputs of the main producing countries. The
ores in question are not in fact directly comparable since some are high-grade ores and
others are not. Part or all of this manganese
ore is exported according to the development
of the local steel and ferroalloy industries.
Table 3 gives the ore import figures for the
main non-producing countries (1977 was an
exceptionally low year). Ore imports have
decreased for certain countries, particularly the United States, having been replaced
by ferromanganese imports.

ORE PRODUCTION

Mining has ceased to be a small-scale
operation, now that manganese has become
a bulk commodity which requires bulk
transport to its markets. An infrastructure
has to be provided such as railroads and ports;
consequently a deposit must be substantial
enough to permit mining at a rate which
amortizes this infrastructure. All the new
mines which have come on stream since 1957
have adopted methods practised in iron ore
mining operations and have annual capacities
of over 1 million tonnes. Where the infrastructure already exists, smaller-scale mining
operations can be expected to be economical.
At present only seven producers are rated at
over half a million tonnes annual production,
and so the industry is highly concentrated.
Since 1970 many small manganese mines
have disappeared either through economic
difficulties or exhaustion of reserves.

The main producer is the USSR, in which
large deposits are located in Ukraine and

Georgia. The ore which is mined has to be
beneficiated. In the Nikopol area, the production of concentrates over 38%Mn exceeds
6 million tonnes per annum; the Chiatura
deposit in Georgia supplies slightly more than
2 million tonnes of concentrates per year.

South Africa comes second with an annual
production of over 5 million tonnes, its principal deposits being located in the region of
Postmasburg and Kuruman Kalahari, a thousand kilometres inland from the nearest ports.
Today, part of the mining is underground and
the major part of the production is controlled
by two companies.

The largest deposit of high-grade manganese
ore is found in Moanda near Franceville in
Gabon, where mining started in 1962. The ore
is evacuated via the port of Pointe Noire in
the Republic of Congo. In a few years' time,
with the completion of the Transgabonese
railroad currently being built, a second transport route will be provided. Mining is at a
yearly rate of over 2 million tonnes, and ambitious production expansion programmes are
contemplated, to be implemented when the
transport capacity becomes adequate.

In Australia only one manganese mine is in
operation — that of Groote Eylandt situated
on an island in the Gulf of Carpentaria on the
country's northern coast. This deposit was
discovered in 1960 and the mine was opened
in 1966. Production reaches 2 million tonnes
per year, a large proportion of the ore being
beneficiated.

Brazil has been a leading supplier in recent
years from the Amapa mine in the Serra do
Navio in the north of the country. The
opening of the mine was in 1957 and by now
the reserves are limited; the export figures are
kept below 1·2 million tonnes per annum.

India was once one of the main manganese
ore suppliers but now its reserves are depleted
and no metallurgical grade is exported. Nevertheless India is still a large supplier of ferruginous manganese ore.

Mexico is a newcomer to the market with
an ore sold in the form of nodules agglomer-
ated in a rotary kiln. The mines are located in
the region of Molango.

In the present situation a single large manganese mine accounts for more than 10% of
the world's manganese ore production and
this will remain so in the future. As already
mentioned, few large deposits are known and

Table 4   World Mn ore reserves
and identified resources (10^3 t Mn);
source US Bureau of Mines

	Reserves	Resources	Total
North America:			
United States	—	66 775	66 775
Canada	—	15 872	15 872
Mexico	1 995	11 428	13 423
total North America	1 995	94 075	96 070
South America:			
Bolivia	—	4 535	4 535
Brazil	39 635	23 582	63 217
Chile	408	—	408
total South America	40 043	28 117	68 160
Europe:			
Bulgaria	4 081	—	4 081
Greece	208	—	208
Hungary	244	—	244
Romania	299	—	299
USSR	680 250	544 200	1 224 450
total Europe	685 082	544 200	1 229 282
Africa:			
Gabon	90 700	—	90 700
Ghana	317	9 070	9 387
Ivory Coast	—	1 360	1 360
Morocco	680	—	680
Rep. South Africa	816 300	707 460	1 523 760
Upper Volta	—	9 070	9 070
Zaire	3 083	997	4 080
total Africa	911 080	727 957	1 639 037
Asia:			
China	15 419	15 419	30 838
India	25 396	10 884	36 280
Japan	861	—	861
Thailand	1 179	2 539	3 718
total Asia	42 855	28 842	71 697
Oceania:			
Australia	145 120	13 605	158 725
Fiji	72	272	344
New Hebrides	72	—	72
total Oceania	145 264	13 877	159 141
TOTAL WORLD	1 826 319	1 437 068	3 263 387

no significant discoveries have been made since 1960. The time lapse between finding an initial indication of manganese ore and a mine becoming operational is generally more than ten years.

Most manganese deposits have been fruitful discoveries, but large stratified manganese oxide deposits are not responsive to modern geophysical techniques. The classical prospecting methods have to be employed. Furthermore, as manganese is not an expensive metal, few systematic programmes of exploration for manganese ore are financed by companies.

ORE RESERVES

According to the US Department of the Interior 'mineral reserves are that portion of mineral resources that have actually been identified and can be legally and economically extracted at the time of determination'. However, there is a lack of agreed definition as to what are 'reserves', 'proven reserves', 'inferred reserves', and 'resources'. Moreover, reserves and resources should be classified according to their manganese content. Most of the ores are used in pyrometallurgical processes to produce ferroalloys and it is necessary to have a manganese content in the burden which is high enough to avoid excessive slag volume. Another requirement is a Mn/Fe ratio of 8 in order to be able to attain the required manganese content in the ferroalloy (73 – 78%). Generally speaking those which contain more than 40%Mn are considered metallurgical ore.

The generally accepted data for manganese ore reserves and resources are those published by the US Bureau of Mines; these are given in Table 4.[1] Recently, however, at the IISI Conference held in Colorado Springs much lower reserve figures were cited from a study carried out by R.L. L'Esperance (see Table 5).[2] The authors take an exacting view of the problem and only admit into consideration deposits which are judged large enough to be exploited in view of the necessary infrastructure and where the reserves are really proven. The figures for 'proven' reserves certainly constitute a slightly more pessimistic estimate than the actual reserves. Part of the 'reserves', since they are in fact extrapolations of known deposits already being mined, will actually become reserves when they have been properly explored.

A regional breakdown of manganese ore reserves is given in Table 6. The two countries whose reserves are most talked about are Russia and South Africa.

In Russia the ore is low grade and needs to be beneficiated. Little published information is available concerning the two principal deposits; figures vary from 306 million to 720 million tonnes manganese content, the higher of these including inferred reserves. Resources are estimated at 300 – 540 million tonnes manganese content. According to general opinion, this country's reserves have been overestimated through lack of information and ignorance of the difficulties met with when trying to beneficiate certain inferior grades of ore.

Table 5   World Mn ore reserves according to L'Esperance

Producer country	Proven reserves, 10^3 t Mn	Resources, 10^3 t ore
USSR	306	1 740
South Africa	249	3 810
Gabon	57	377
Australia	33	300
Ghana	1	30
Brazil	11	180
India	3	114
World total	678	6 935

Table 6   Distribution of world Mn ore reserves (two assessments)

Country	After US Bureau of Mines, %	After L'Esperance (Table 5), %
Rep. S. Africa	44·7	36·7
USSR	37·2	45·1
Gabon	5·0	8·4
Australia	7·9	4·8
India	1·4	0·4
Brazil	2·2	1·6
Others	1·6	3

In South Africa the problem is similar. A high proportion of the mining is at considerable depth underground. According to the South African Minerals Bureau,[3] it has been demonstrated with the Wessels mine that it is possible to mine manganese ore economically to a depth of 300 m. Based on that assumption, the reserves (mostly not proven but inferred) have been fixed at 12 billion tonnes of ore, that is 3 billion tonnes manganese content.

According to the US Bureau of Mines, reserves amount to 816 million tonnes and resources to 707 million tonnes. L'Esperance's estimate is 249 million tonnes manganese content for proven reserves and 3·8 billion tonnes of ore for the resources. Assuming a 4% growth rate for manganese consumption, the actual reserves are sufficient to meet world requirements for the next 55 years according to the US Bureau of Mines data, but for only 29 years if L'Esperance's more pessimistic data are to be used as a basis. In view of the fact that a large proportion of the resources of countries such as South Africa and Gabon will progressively become proven reserves, one may assume that even under the worst conditions manganese needs may be assured for the next 50 years by currently known deposits. It is difficult to foresee what new discoveries may be made in the meantime, but it is reasonable to expect new deposits to be found despite the difficulties met with in prospecting for this metal.

The problem with manganese is not a lack of reserves but the extreme concentration of this industry. It is obvious that if for one reason or another one of the existing main producers was no longer in a position to produce, then the whole market would be severely perturbed. The majority of the other mines could increase production but only after a certain time lapse to allow for study and the provision of the extra equipment which would in most cases be indispensible.

Within the next 15 years or so a new source of manganese may become accessible, namely the mining of nodules on the ocean floor.

Deep-sea nodules

A considerable exploration effort has been made over the past ten years to assess the reserves of nodules containing metallic elements which lie on the ocean floor. These nodules, which are continuously formed from the minerals in solution in sea water, contain 8 – 40%Mn, with an average value near to 30%. They are a potentially important future source of copper, nickel, and cobalt; manganese, albeit the most substantial, is of secondary interest. The figures put on reserves of nodules vary greatly according to whether or not one counts deposits actually identified or makes extrapolations to the whole surface of the ocean floor, and according to the minimum density per square metre regarded as necessary for an industrial operation.

Moreover, it is difficult to assess what proportion of the nodules it will be technologically possible to recover. Recovery rates better than 50% are not thought feasible until after the year 2000, and a first industrial operation will probably recover only 20–25% of the nodules.

A. A. Archer of the Institute of Geological Sciences[4] arrives at recoverable reserves of either 16·7 or 49·7 billion dry tonnes, the former from discovered locations, the latter from the ocean floor as a whole. A. F. Holster of the Ocean Mining Administration[5] puts these figures at either 29·4 or 69·4 billion tonnes. On the other hand, assessments made by Lenoble and Rogel[6] put the quantity of nodules which can be considered as ore between 1 and 3 billion tonnes. If one assumes an average manganese content of 29%, the reserves of manganese which these nodules offer would lie between 260 million and 18 billion tonnes according to the more optimistic estimates.

In the short term, it is clear, manganese from nodules will have no decisive advantage over the major land reserves. Most of the international consortia involved in developing techniques for recovery and processing of deep-sea nodules are not even thinking of recovering manganese in the first stages of industrial operations. In the longer term, manganese from the sea bed will probably help towards supplying part of the need for this metal.

CONCLUSIONS

The known manganese resources will be able to meet the market's needs for the next 50 years. To these resources will be added within the next 10–20 years manganese extracted from sea-bed nodules.

Manganese therefore remains an abundant and relatively inexpensive element. Its use as an alloying element in the steel industry should develop when, as a result of appropriate research efforts, all the possibilities of this element are fully realized. Outside of the steel industry, particularly in the field of corrosion resistance, manganese could find a market opening, either being added to or replacing other metals with more limited reserves.

Large manganese deposits on the earth's crust, situated rather deeper in the ground, certainly remain to be discovered but unfortunately effective physical methods of detecting them are not at present available and prospecting for a metal which is both cheap and abundant does not arouse really great effort.

The only problem concerning the steel industry's manganese supply is linked to the restricted number of known deposits and the concentration of a high proportion of the free world's reserves in a single country. A catastrophe in one of the large mines or a political change in one of the producing countries could have serious repercussions on the market's supply. Nevertheless, whatever happens, it is difficult to imagine that a source of supply would be cut off for very long.

REFERENCES

1 US Bureau of Mines publication 'Manganese', 1977
2 P. HAWKINS: 'Deep-sea nodules — an essential new source of manganese,' IISI, 12th Annual Meeting and Conference, Colorado Springs, USA, October 1978
3 W. A. DUKE: *Metals and Materials*, December 78/ January 79, 17
4 A. A. ARCHER: Document IGS, 1589, 200 4176, Inst. Geological Sciences, London
5 A. F. HOLSTER: 'Manganese nodules — resources and mine site availability', Ocean Mining Administration
6 J. P. LENOBLE and P. ROGEL: 'Polymetallic nodules', *Annales des Mines*, May 1977, no. 5, 73-81

R. B. Nicholson and P. G. Cranfield
Nickel

In order to make an assessment of future nickel consumption and to comment on future growth, a fairly detailed analysis of past developments is necessary.

NICKEL DEMAND

Nickel consumption was already growing rapidly in the early years of this century from less than 10 000 metric tons to probably around 25–30 thousand tons annually during the first World War, when the majority of nickel was used in armaments. A serious slump followed the war, as the demand for armaments disappeared and large quantities of these were appearing in the form of nickel-

---

The growth of nickel consumption from the beginning of the present century is reviewed and its close relation to stainless steel production is indicated. While growth slowed during the 1970s in Europe, the USA, and Japan, there is still considerable potential for increased nickel consumption in developing countries and from new applications in the developed economies. There is no intrinsic shortage of nickel in the earth's crust and shortages of supply are likely to occur only if the price of nickel fails to reflect the capital cost of new capacity.

The authors are with Inco Europe Ltd, London.

Table 1   Nickel consumption (primary) in non-communist world

Usage	%
Stainless steel	45
Other alloyed steels	10
Non-ferrous alloys	20
Plating	10–15
Castings	5–10
Other (chemical, batteries, etc.)	5

containing scrap. Added to this were the effects of the serious world-wide recession in the early nineteen-twenties. During the 1920s and '30s, largely as a result of the work by the International Nickel Company and Mond Nickel Company, new applications for nickel were developed and existing uses expanded. By the mid-'30s the consumption of nickel had grown to around 75 000 tons. Major uses were nickel-base alloys, constructional steels, cast iron and steels, copper–nickel alloys, permanent magnets, and electroplating.

However, the present pattern of nickel consumption was established after the second World War. Nickel is rarely used in the un-alloyed state except for electroplating or wrought pure nickel mainly for chemical, electronic or coinage applications. Today the consumption of nickel in the non-communist world is broadly as shown in Table 1.

Table 2   Growth rates in non-communist world

	Average annual growth rate, %		
	1950-60	1960-70	1970-78
Nickel consumption	5·1	8·0	1·9
Stainless steel production	7·1	7·9	3·8
Index of industrial production	4·3	5·7	3·0

Stainless steel accounts for nearly one-half of nickel consumption and, as the tables below will show, has been the fastest growing major use for nickel since its growth has exceeded that of nickel. The future for nickel is clearly closely linked to that of stainless steel. The growth in consumption of both nickel and stainless steel has been rapid in the post-war years.

The following factors are important in determining the rate of increase in the consumption of a metal:

1   the rate of economic growth, or more particularly the rate of growth in those sectors of the economy in which the metal is predominantly used

2   the rate of substitution in existing applications by competing materials or, on the positive side, the substituting for other materials

3   in the case of alloying elements, downgrading or up-grading in the alloys

4   the development, if any, of major new applications.

Between 1946 and 1977, nickel consumption in the non-communist world grew at an average rate of 5·7% per annum, during which period the growth rate varied considerably. Table 2 shows the trend of growth in nickel consumption over the three periods 1950-60, 1960-70, and 1970-78. This is compared with the growth of stainless steel ingot production in the non-communist world over the same period. The growth in the index of industrial production for all OECD member countries combined, over the same period, is a reasonably good measure of the growth in economic activity in the mining and manufacturing sectors.

It is clear from these rates of growth that the 1960s were an exceptional period in terms of all three rates: nickel, stainless steel, and economic; while the 1970s saw a serious reduction in growth in all three. The reasons for the slowdown in economic growth are many, but certainly a major cause was the quadrupling of the oil price from the end of 1973. However, it is doubtful whether the high growth rates of the 1960s and the world-wide boom of 1973 could have been sustained even without the oil crisis. The reality today is that the major economic goal of nearly every government is not that of economic growth but the more urgent need to contain the high rates of inflation. The reduction in the rate of economic growth has been a major cause of the low growth in nickel consumption in the '70s.

Stainless steel production still grew at a slightly faster rate than the world economy, which helped to boost nickel. The relatively poor performance of nickel in the last few years is largely a result of the lack of capital expenditure, and particularly the low level of investment in those sectors where nickel is used. The investment sector accounts for some two-thirds of total nickel consumption. The economic recovery since the 1975 recession has largely been a recovery of the consumer and automotive sectors. Low levels of capacity utilization, and doubts concerning future levels of demand, are probably two of the more important factors behind the lack of an investment boom, although escalating costs and environmental controls also play their parts.

Other uses for nickel are more directly related to the capital goods sector and showed little growth during this period. Nickel consumption in alloy steels for example in 1977 was only marginally above that of 1970.

Table 3   Post-war nickel consumption and
stainless steel production (10^3 tonne)

	1950	1960	1970	1978 (est.)
NICKEL CONSUMPTION				
W. Europe	30	95	175	200
USA	90	100	150	175
Japan	1	20	100	105
Others	4	5	25	45
total non-communist world	125	220	450	525
STAINLESS STEEL INGOT PRODUCTION				
W. Europe	224	960	1 970	2 720
USA	755	908	1 158	1 760
Japan	4	238	1 643	1 940
Others	25	44	179	250
total non-communist world	1 008	2 150	4 950	6 670

Similarly the foundry sector showed no over-all growth over the period despite high levels of demand during the 1973–75 period when investment peaked. The same general pattern was true of the nickel-base alloys although the underlying rate of growth was stronger. Nickel for electroplating increased quite rapidly in the early years of the decade but has suffered from substitution by plastics in automotive trim in recent years.

Certainly the growth rate in nickel consumption is declining but probably not to the extent indicated by the last few years.

GEOGRAPHICAL DISTRIBUTION

Before reviewing predictions of future demand for nickel it would be useful to examine the geographical division of nickel consumption and stainless steel production in the post-war period (Table 3).

The 1960s were an exceptional period for both nickel and stainless steel throughout the world. Not only was economic growth rapid but it was particularly strong in those sectors using substantial quantities of nickel, e.g. investment in chemical and petrochemical plant. There was also a very strong substitution factor favouring stainless steel particularly

in Europe. Kitchen sinks, pots and pans, hollow-ware, washing machine drums, and cutlery went from enamel-ware etc. to stainless steel. This movement was so strong that within a few years in many European countries over 90% of all kitchen sinks were produced in stainless steel and future potential growth is thus limited.

It can be seen that in terms of both nickel and stainless steel the USA was by 1950 already at a high level relative to the rest of the world. Consumption growth in the USA between 1960 and '70 was still at a reasonably high rate but well below that of the rest of the world, but between 1970 and '78 it has exceeded that of Europe and Japan. This reflects the relative growth of the US economy and the import quotas on stainless steel, but also that in 1970 the US was in recession and suffering from the end of a nickel shortage.

Nickel consumption in Europe grew much more rapidly during the reconstruction period of the '50s and the high growth and substitution period of the '60s. Growth in the last eight years has been low, partly because of the poor economic situation in Europe (particularly since 1975) but also due to a slowing down as some applications at least, such as sinks, approach saturation level.

The pattern of growth in Japan, where

81

most of the nickel is consumed in stainless steel, is also interesting. The startling growth between 1960 and '70 was mainly the result of rapid increase in the domestic demand for stainless steel from around 150 000 tons in 1960 to around 850 000 tons in 1970. Exports of stainless steel also rose from around 20 000 tons in 1960 to nearly 300 000 tons. Since 1970 the upturn in Japanese stainless production has been the result of increased exports which exceeded 600 000 tons in 1977. The domestic market has not, in fact, expanded; instead, the same slowing down has occurred in Japan as in Europe.

The interesting development is in the rest of the world, outside the three major developed areas. Nickel consumption in these diverse areas nearly doubled from 25 000 tons in 1970 to 45 000 tons in 1978. In absolute terms these countries are now consuming nearly half as much nickel as Japan. Some of the countries under this heading, such as Canada, South Africa and Australia, are already developed economies but still have considerable growth potential. Also included are areas where nickel consumption is growing rapidly, e.g. Brazil, India, South Korea, and Taiwan.

THE FUTURE FOR NICKEL

What is the most likely increase in nickel consumption during the next 20 years? Europe, the USA, and Japan together account for over 90% of nickel consumption in the non-communist world. Growth will largely depend on the four main determinants mentioned above:

economic growth
rate of substitution
down-grading
new applications

An entire conference could be devoted to the anticipated economic growth in the developed economies and still no satisfactory answer would evolve. The most prudent assumption might be that growth will not attain that of the nineteen-sixties but could exceed that of the mid-seventies.

In the case of substitution there are no indications of any major movement either towards nickel or away from it to competing materials. Certainly bright trim in the automotive sector has been replaced in many cases by plastics or painted surfaces, but this is due rather to fashion and fuel conservation measures than to cost savings or any better performance by the competing material.

In the case of downgrading of nickel in alloys, again there is little evidence of this on a significant scale. The threat of nickel-free stainless steels, which was discussed for many years, has never materialized. Austenitic stainless steels have remained at a stable 70% of world production for several years. The nickel-free grades continue to be used in the traditional consumer applications and the high-chrome molybdenum steels have made some inroads in very specific applications, but these latter grades are difficult to produce, contain costly alloying elements, and are likely to remain a small tonnage business.

As far as major new applications are concerned, these are by their very nature almost impossible to predict over the long term, but one should cite the possibility that batteries for electric vehicles could well develop into an important market in future years.

The future importance of the developing world in terms of nickel consumption is equally difficult to predict. The economist W. Leontief in his study for the United Nations entitled 'The Future of the World Economy' pointed out that in 1970 the developing nations accounted for 14% of non-communist world GDP. On the most optimistic assumption of high economic growth in these areas, they could account for as much as 30% of world GDP by the year 2000. However, Leontief himself states:

'The future can rarely be predicted with precision; and the future of such a complex phenomenon as the world economy is particularly difficult to anticipate or even to visualize'.

Having made this considerable qualification, Leontief gives the possible scenario for economic growth and nickel consumption from 1970 to 2000 shown as Table 4.

In contrast, a more pessimistic outlook is that of W. Malenbaum in his study 'World Demand for Raw Materials in 1985 and 2000'.

Table 4   Annual growth rates 1970–2000
according to Leontief (%)

	Developed market economies	Developed centrally planned economies	Developing economies
GDP	3·6	5·1	7·2
Nickel consumption	3·6	5·4	7·1

Table 5   Average annual growth rates
according to Malenbaum (%)

	1951–75	1975–2000
World GDP (including USSR, E. Europe, China)	4·7	3·3
World population	2·0	1·8
World GDP per-capita	2·7	1·5

His economic projections are very much more pessimistic (see Table 5). Malenbaum develops the idea of the intensity of use for a metal, which is the weight consumed per dollar of GDP (constant prices), and observes that the intensity of use for most metals has been declining for two reasons: (1) shifts in the types of final goods and services that world consumers and investors demand; (2) technological developments which alter the efficiency with which raw materials are consumed in the production of final goods. Certainly, both points are valid. In a fast-growing sector such as data processing equipment the value and contribution to GDP is relatively high but the consumption of metals is low. Also it is true that in due course better design leads to the use of thinner gauges of metal goods, for example.

In the case of nickel, Malenbaum's 'intensity of use' shows an apparent steep decline in the period 1971–75. As mentioned earlier, it would be wrong to interpret this entirely as a structural decline in the demand for nickel. It was partly a cyclical problem in the sense that in the current economic cycle the investment sector has been comparatively weak. Malenbaum's forecast of worldwide nickel consumption growing at 2·8% per annum between 1975 and 2000 is certainly at the lower end of the range of forecasts which have been published.

To summarize, it may be prudent to assume that economic growth and nickel consumption in the developed economies will improve on the recent growth rates, although by how much is questionable. In developing countries, once a certain level of industrialization has been reached nickel consumption will increase rapidly. In the case of China, nickel production is currently very low and any short-to-medium term increase in consumption could represent a demand for Western supplies. One further point which should be emphasized is that superimposed on trend growth of consumption is the cyclical growth which follows the business cycle. The cyclical fluctuations in nickel consumption greatly exceed that of the business cycle; this is due to stock changes throughout the whole chain of raw material to finished product, and to the influence of the scrap cycle.

NICKEL RESERVES

The estimates of nickel reserves vary slightly according to source, but US Bureau of Mines data show 'proven and probable' land-based nickel reserves to be as follows (for 1976, in millions of tonnes):

Cuba	18
New Caledonia	14
Soviet Union	12
Canada	9
Indonesia	7
Africa (minimum)	2
Australia	5
Others	7
Contingency	1
	75

One can assume a static level of production in order to calculate the life of the reserves. If one divides the latest assessment of reserves by 1976 nickel production, the life of econo-mically recoverable land-based reserves is 105 years; and if one includes potential land-based reserves this is extended to about 250 years. Production is, of course, increasing over time in line with growth of demand. On the other hand, estimates of reserves also increase over time as metal prices increase, mining and processing techniques improve, and new ore bodies are found. The estimates of total land-based nickel ore reserves, proven and probable, have increased as follows in recent years:

Contained nickel, 10^6 tonne

1950	1964	1974	1976
14	28	33	75

Throughout the period 1946–77 the calcu-lation of the life of land-based nickel reserves has always been in the range 100–150 years.

In addition to the land-based reserves there exist vast reserves of nickel-bearing manganese nodules which may start to be mined by the year 2000. Total nodules contained in the world's oceans are thought to be of the order of 1.5×10^{12} tonnes, containing thousands of millions of tonnes of nickel although only a proportion would be economically recoverable.

The US Ocean Mining administration esti-mated that there could be 180–460 potential first-generation production plants. These were defined as an area with a minimum density of 10 kg of manganese nodules per m^2, a metal content of copper plus nickel between 2·25 and 4%, and a marginal cut-off grade of copper, nickel, cobalt totalling 2%. The first-generation plants are further defined as having an annual output of 3 million t over a life of 25 years, i.e. a total of 75 million t containing almost one million t of nickel. Grossing up by the 180–460 potential plants would result in a potential recovery of 180–460 million t of nickel.

Nickel is certainly not one of the metals about which there are any doubts of sufficient reserves until at least well into the next century. However, there may well be periods of temporary shortage and consequent higher prices. The cost of building a new mine/smelter/refinery will vary considerably, but an average figure would be about US$10 per lb of nickel annual capacity. Such an operation, to show a profit, would need a nickel price of at least $3 per lb. During the last two years the price has hovered around, or more often below, the $2 level. As a result, apart from the expansions already under way, very little additional capacity is planned for beyond 1985. The huge Gag Island project in Indonesia with a planned capacity of about 50 000 t of nickel per annum by the late '80s was recently cancelled and current non-communist world nickel production is only about 65–70% of the level of capacity.

CONCLUSIONS

It is reasonable to conclude that, unlike the situation with some metals being discussed at this conference, there is no shortage of available nickel resources in the earth's crust. Shortages of supply are likely to occur only if the price of nickel fails to reflect the very considerable capital costs of adding new mining, smelting and refining capacity, espe-cially in the more remote regions of the world.

Growth in demand for nickel is unlikely to return to the exceptionally high levels of the 1960s. Future growth will depend on the development of new markets in the industria-lized world and the rate at which the less industrialized countries start to consume nickel in its traditional applications, especially stainless steel.

A. M. Sage
Vanadium

The first commercial use of vanadium was for the production of certain grades of tool steels. These steels made use of the refractory characteristics of vanadium carbides present in the steels and enabled the tips of the tools, which attain a high temperature, to maintain their effectiveness during the high-speed cutting operation. Vanadium steels were also adopted for dies used for the production of hot forgings where the refractory carbides provide wear resistance at the high temperatures attained during forging and pressing. Later, vanadium steels were specified for certain components in steam power plants,

particularly for rotors, steam pipes, superheater tubes, and turbine casings and also for rotors and blades in some designs of gas turbines. In all these applications, precipitates of vanadium carbides enabled the steels to maintain their strength and to resist creep at the operating temperatures. At an early date vanadium was also added to forging steels to prevent grain growth at forging temperatures and thus to ensure a fine grain size in the finished product. These applications remained the most important uses of vanadium until the mid-1960s, and accounted for most of the world demand of about 40×10^6 lb (18×10^6 kg) V_2O_5 which had developed up to this date.

A brief account of the growing use of vanadium in steels and alloys and as a catalyst is given. It is considered that the past expansion of vanadium consumption will continue, with the probability of new applications to augment this upward trend. The current sources of vanadium are detailed, showing that reserves are likely to be ample for several centuries to come, and that supplies are less vulnerable to political crises than are many other minerals.

The author is with Highveld Steel and Vanadium Corporation Ltd, London.

In the 1960s the production of microalloyed steels, which had been developing steadily since the end of the war, experienced a major growth largely (though not entirely) due to the development of high-strength pipelines to carry gas and oil over long distances. As a result of this the world demand grew rapidly over a period of three to four years and attained a level of approximately 65×10^6 lb (29×10^6 kg) V_2O_5 by 1970. More recently vanadium has been used for heavy-duty rail steels and high-strength cold pressing steels for automobile parts and the use of vanadium for both of these applications is likely to grow in the next decade.

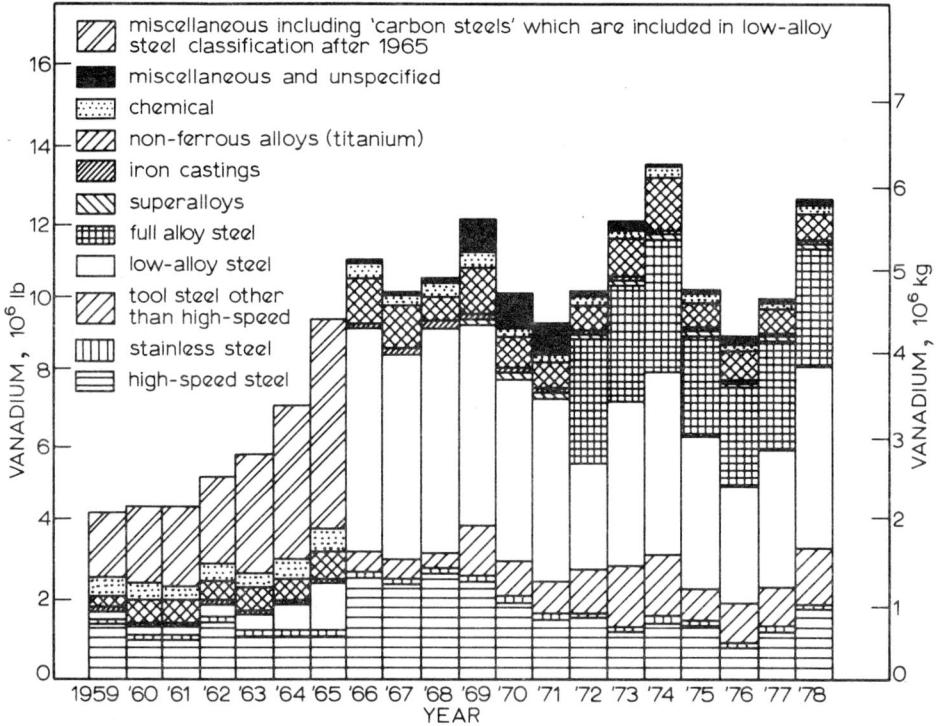

1   Vanadium consumption in USA 1959–78 by end use

Other uses of vanadium include titanium alloys and chemicals. Vanadium is added to titanium alloys which are required to operate at high temperatures and such alloys are used in many aircraft turbine components and also in certain parts of the airframes of jet aircraft which attain a high temperature during flight. Parts of the doors of the undercarriage of the Concorde, for example, are made of a titanium alloy containing 4% vanadium. This use of vanadium accounts for approximately 10% of the vanadium consumption in the USA, which in turn accounts for approximately one-third of the vanadium consumption in the world. Vanadium chemicals have been used for catalysts in sulphuric acid manufacture for many years and more recently new types of vanadium catalysts have been developed for removing sulphur from effluent gases from power stations. The use of vanadium for these chemical purposes

accounts for up to 10% of the world's vanadium production.

Vanadium has been used in metallic form in the cans of atomic reactor vessels on account of its low cross-section absorption of neutrons and it was used for this purpose in the original Dounreay breeder-reactor. At present conditions do not favour the use of vanadium but conditions could change and this market always remains as a potential for vanadium.

Actual figures for the end use and consumption of vanadium are difficult to obtain except in the USA where detailed figures are published annually by the US Bureau of Mines (Fig. 1). It is possible, however, to calculate the total consumption of vanadium in most countries of the non-communist world from statistics of imports and exports and it is estimated that in 1979 a possible

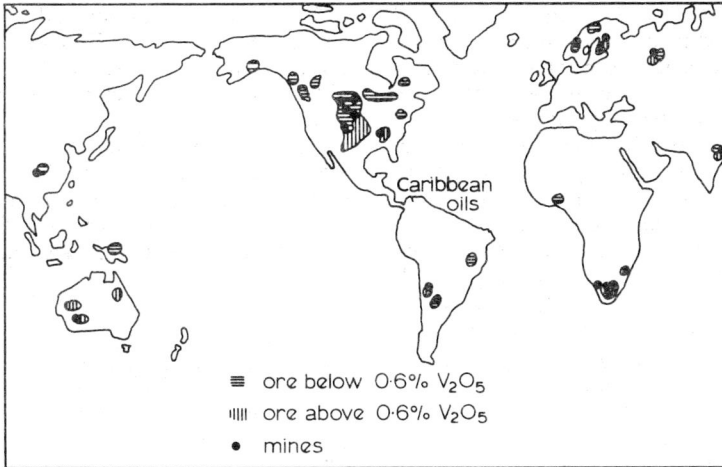

2   World sources of vanadium

all-time maximum consumption of more than 80×10^6 lb ($36 \cdot 3 \times 10^6$ kg) V_2O_5 may be achieved. Statistics indicate that over the last 15 years the non-communist world market has been growing at an average rate of about $2 \cdot 5\%$ and if this continues to the year 2000 it will mean that by then the consumption will be at the rate of over 130×10^6 lb (59×10^6 kg) V_2O_5 per annum.

FUTURE MARKETS

No-one can be sure these days how any metal market will develop because the technical requirements of the ultimate user can change rapidly and unpredictably and the relative price and availability of would-be alternatives can vary as quickly. It is, however, considered that the use of vanadium in alloy steels, including tool and die steels and high-temperature steels, will grow with the world's aggregated gross national products roughly in line with steel production. On the other hand the use of vanadium as a microalloying element, especially in pipelines, while subject to many factors and wild fluctuations from year to year will probably on average show greater growth. Newer uses for vanadium which could lead to further overall growth include steels used in forged crankshafts which replace

heat-treated carbon steels, cold pressing high-strength steels used to reduce the weight of automobiles, and heavy-duty rail steels installed in special railways where axle loads are high. Other possible uses include tools made from new alloy steel compositions by powder metallurgy.

If molybdenum remains in short supply new vanadium-bearing steels may also be developed for heat-treated engineering steels and possibly other steels which at present contain molybdenum; some estimates of the future demand for vanadium suggest that an overall average growth rate of 4% or more will be achieved.

CURRENT SOURCES
IN NON-COMMUNIST WORLD

Vanadium is one of the most plentiful metals in relation to its demand and, as will be seen from the world map in Fig. 2, it is one of the most widely distributed. It always occurs with other minerals and is frequently produced as a co-product or by-product, but even in the richest ores it is rarely present in amounts as high as 2% V_2O_5 equivalent. It is most commonly found in magnetites, which often also contain titanium and in some cases, though not always, the iron or the iron and titanium are produced as co-products. Other major mineral deposits from which vanadium is extracted include phosphate ores from which the phosphorus is produced as ferrophosphorus, uranium ores where the uranium is extracted as a co-product, and oils and tar sands from which petroleum products are obtained. An important source in the USA is the alumina clay of Arkansas from which vanadium is extracted as a primary product.

In the past it was extracted from asphaltes where it exists as vanadium sulphide (such deposits in Peru were once the major source) although these are no longer worked for vanadium. It was also extracted from lead and zinc ores of South-west Africa but this production is at present in abeyance. Vanadium has been extracted from bauxite residues in the production of aluminium. All of these sources are however out of production at least for the time being.

Although, as indicated in Fig. 2, vanadium is widely distributed over the world's crust, today most of the vanadium used by the non-communist world comes from South Africa, the USA, or Scandinavia.

South Africa
South Africa currently produces the largest quantity of vanadium in the non-communist world. It is produced at three major mines, in two of which V_2O_5 is the sole product, while in the third and largest the vanadium is supplied to the market mostly as a vanadium-bearing slag which is a co-product in the production of iron and steel supplied to the domestic market. In all the mines vanadium occurs in a titaniferous magnetite in the Bushveld complex where reserves are estimated to be well over 68 000 million lb V_2O_5.

Scandinavia
Vanadium is produced in Finland where it is obtained as a co-product with iron and titanium. The chemical composition of the ore is similar to that found in South Africa, but the vanadium is present in an occluded form instead of being chemically combined. This means that the titanium can be separated by crushing and flotation and supplied to the pigment and welding flux industries economically.

A new mine was opened at Mustavara in Finland in 1977 and if operated at the full rated capacity without further development will supply the market at least until the year 2000.

There is a smaller production from a similar ore in Norway where vanadium is produced as a slag during ironmaking and converted directly into a special grade of ferrovanadium which is supplied to the steel industry.

USA
Vanadium is supplied from a variety of sources and ores in the USA:

(a) *Uranium ores of Colorado and Utah.* The V_2O_5 is produced from these ores as a co-product with uranium. In recent years several new mines have been opened and it is expected that they will continue to produce V_2O_5 at the current rate of 9×10^6 lb until at least the year 2000. If certain environmental problems can be overcome this production will probably be increased.

(b) *Phosphate ores of Wyoming.* V_2O_5 is produced from these ores as a by-product in the production of ferrophosphorus. This is a relatively small source, but the present production rates of about 3×10^6 lb V_2O_5 are expected to keep going until the year 2000 and the reserves would support a much greater rate of extraction.

(c) *Alumina clays of Arkansas.* V_2O_5 and other oxides of vanadium are produced from this ore as the sole product of the mine at Wilson Springs; these are converted to vanadium carbide and vanadium nitride, which are used as alternatives to ferrovanadium in steel production. This plant has operated intermittently in the last five years, but it is capable of producing about 15×10^6 lb V_2O_5 per year when on full stream. The reserves from this ore amount to 120×10^6 lb V_2O_5 and the plant could therefore keep going at full capacity until 2000.

Caribbean oil residues
These are a significant source of vanadium and they could become a major source in the twenty-first century. At present approximately 4×10^6 lb V_2O_5 probably originates from

Table 1   Reserves of vanadium from current sources
in non-communist world

Country	Proven reserves, 10^6 lb V_2O_5 equivalent	Rate of extraction at max. output, 10^6 lb V_2O_5/year	Date of exhaustion at current extraction rate, year A.D.
South Africa	68 000	54	3238
USA: Colorado 180		9/12·0	2000–2205
Arkansas 120		15	1988
Wyoming 500	800	3	2149
Finland	500	12	2021
Norway	500	1·5	2300
Venezuela	—	—	—
Australia (Barambie)	2 400	5	2460
Chile	1 800	—	—
India	—	—	—
total proved world reserves	72 200	99·5	2700

Based on data from 'Vanadium Supply and Demand Outlook', National Materials
Advisory Board, USA, 1978

them. The vanadium from Caribbean oil reaches the world market via several routes:

(1) Spent nickel–molybdenum catalysts operating in petrochemical plants treating Caribbean oils.

(2) Residues obtained from the cleaning of oil tankers carrying oil from the Caribbean.

(3) Boiler ash and fly ash from power plants, particularly those on the eastern seaboard of the US where Caribbean oils are used.

The volume of V_2O_5 reaching the market from these sources depends on the costs of transporting the ashes and residues to the vanadium plants in relation to the price of vanadium.

At the end of 1979 or early in 1980 a plant is due to come on stream in Venezuela which will up-grade oil and which will give a by-product from which vanadium is to be extracted. It is estimated that this could yield about 1×10^6 lb V_2O_5 in 1980 and some estimates for the ultimate production of V_2O_5 by this route are as high as 10×10^6 lb per year.

Australia
There are considerable reserves of vanadium in North-western and South-western Australia

and in oil shales in the Julius Creek area of Australia. At the present time a plant is under construction for the extraction of vanadium from magnetites in the north of Western Australia. This plant has an expected capacity of 3×10^6 lb/year and will come on stream in 1979–80. There are plans for the capacity of this plant to be increased to $7·5 \times 10^6$ lb/year V_2O_5 later in the '80s.

Chile
Vanadium is present in a low-grade slag produced from a steel plant smelting magnetite ores in Chile. This slag is shipped to the USA where it is mixed with other vanadium-bearing materials for the extraction of vanadium.

India
It is understood that some vanadium is extracted from magnetite in India as V_2O_5 which is converted to ferrovanadium and that all of this is used for the production of vanadium steels in India.

RESERVES
The estimated reserves in those countries currently producing or about to produce vanadium products in the non-communist world are given in Table 1. It will be seen that

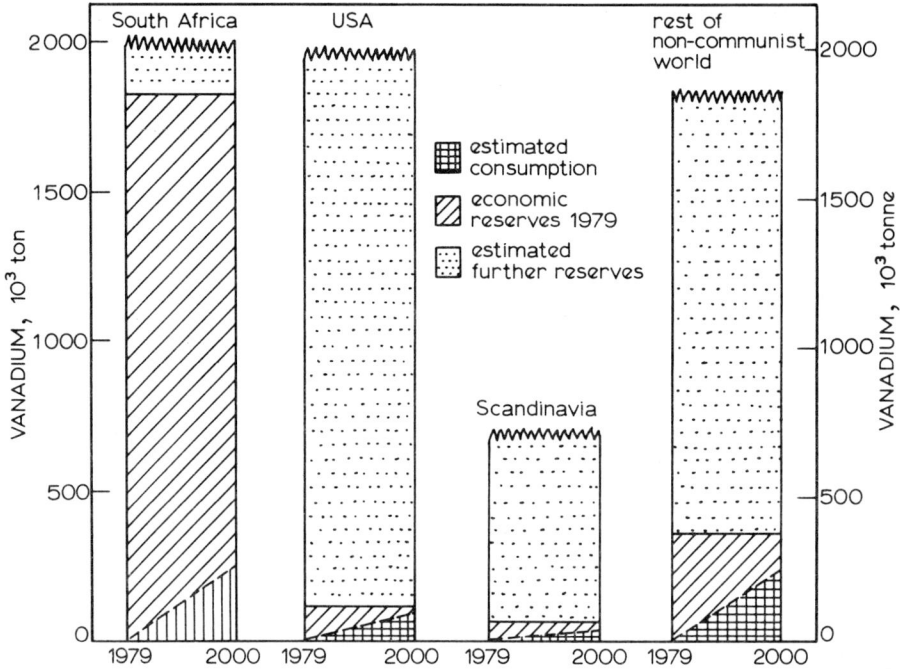

3 Effect of estimated consumption on vanadium reserves in major supply regions

Table 2   Unworked resources of vanadium
in non-communist world (as at 1979)

Country	Ore type	V content, %	Reserves, 10^6 lb V_2O_5
USA:			
Alaska	titaniferous magnetite	0·05–0·02	80 000
Minnesota		0·05–0·11	400
Wyoming		0·02–0·36	1 000
California		0·7	120
New York		0·15–0·3	1 000
Idaho	oil shale	0·4–0·3	1 000
Canada:			
Quebec	titaniferous magnetite	0·1–0·2	12 000
Alberta	tar sands	150 ppm	14 000
	oil	230 ppm	unknown
Australia:			
Windowe	titaniferous magnetite	1·1	600
New Zealand	titaniferous magnetite	0·2	100
India	bauxite	0·1	unknown
Japan	beach sands	—	unknown
Central Africa and Mozambique	titaniferous magnetite	0·3–0·4	4 500

Data from same source as Table 1

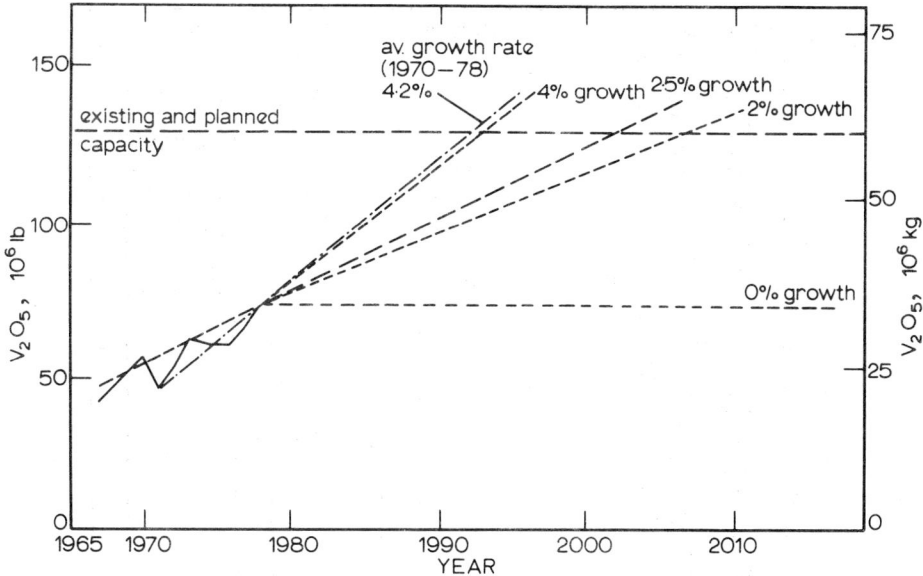

4   Estimated effects of different market growth rates
     on absorption of present non-communist world
     capacity for vanadium production

the total reserves in these regions amount to about 70×10^9 lb V_2O_5 and represent about 700 years' supply at the 1978 rate of mining.

It was indicated above that most of the vanadium used in the non-communist world at present comes from South Africa, the USA, and Scandinavia; the effect of mining at current rates on the reserves in these countries is illustrated in Fig. 3.

It will be appreciated from Fig. 2 that apart from the regions where vanadium is at present extracted there are many other areas in the world where there are vanadium-bearing ores from which vanadium could be produced. The most important of these with estimates of their reserves are given in Table 2. There are also large reserves of vanadium in the communist world. The most important deposits are understood to be the vanadium-bearing magnetites in the Ural mountains of Russia from which vanadium is extracted at an annual rate estimated to be at least 35×10^6 lb V_2O_5. Until about five years ago

some of this was exported to the non-communist world as slag from an ironmaking plant, but at the present time this is used in Russia and in Eastern Europe, the latter in fact tending to be a net importer of vanadium. It is known that there are deposits of a similar ore in Poland and consideration is being given to the exploitation of this source.

China is reported to have large reserves of vanadium minerals and to be planning the construction of plants for the production of V_2O_5 and ferrovanadium. At the present time it imports vanadium from Japan and Europe.

A small proportion of the world's vanadium is recycled. As mentioned above, several million pounds of V_2O_5 in the form of spent catalysts from sulphuric acid plants are recycled through V_2O_5 plants. In the tool steel industry a proportion of scrap high-speed tools are recycled to steel companies and this probably accounts for about 20% of the vanadium used for high-speed tool steel production, or about 2·5% of the vanadium consumed in the non-communist world.

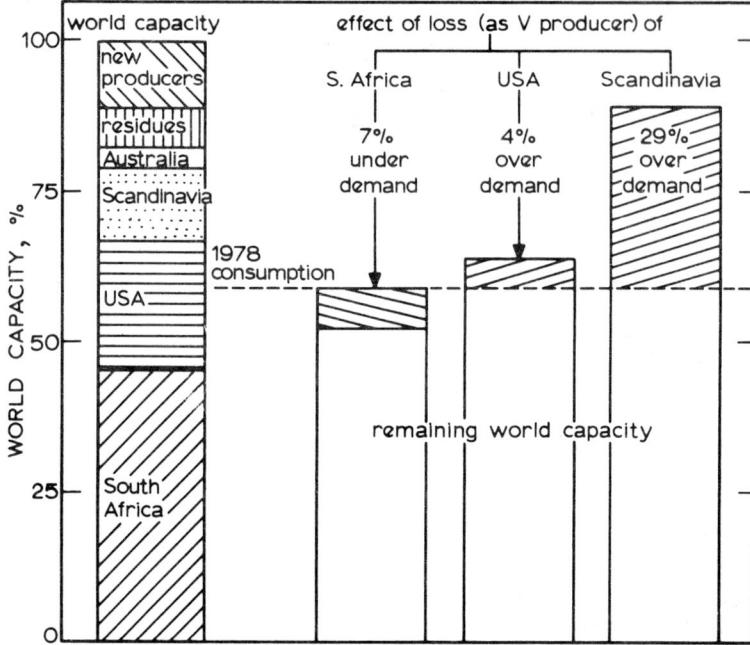

5   Effect of collapse of major sources of vanadium
     on world demand (based on 1978 consumption)

MINING AND PRODUCTION:
THE IMMEDIATE OUTLOOK

It is very difficult to forecast demand for a metal on a regional or global basis as many factors control the rate of growth of a market and under present world trade and economic conditions these factors are particularly uncertain. In an attempt to give a picture of the future production/consumption balance for vanadium up to the year 2000, the demand at several alternative growth rates has been calculated from 1979 through to 2000 and the results are shown in Fig. 4. It will be seen that at an assumed average growth rate of 2·0% the estimated present production capacity will be absorbed by 2007. If, however, an average growth rate of 4% is achieved, this capacity would be taken up by 1993. In the event that the market grows at this rate there will be adequate time for the vanadium

industry to invest in additional capacity to keep ahead of the demand, and, as has been shown, there are certainly adequate sources of raw materials to support such expansion of the industry.

The security of supply is always of importance to users of any basic material and to those engaged on developing new markets for it. This is especially true at the present time and has been a major concern for all users of minerals since the Middle East oil crisis of 1974. It is therefore useful to consider the effect on the total supply situation of any one major source of vanadium going out of production. The estimated effect of an interruption in production in South Africa, or the USA, or Scandinavia, on the basis of 1978 consumption and 1978 capacity plus planned additional

92

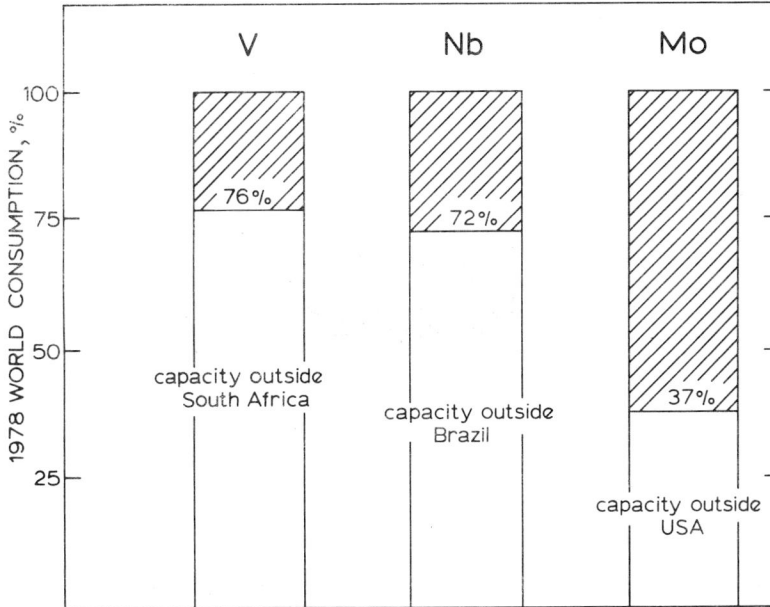

Based on data from
1   'Vanadium' by R. H. Herbertson and A. M. Sage
      Met. Bull. Alloy Conf. Zurich 1977
2   'Columbium', US Bureau of Mines 1978
3   'Molybdenum', US Bureau of Mines 1979

6   Effect of loss of major alloy supply region on demand
     (based on 1978 consumption and capacity of V, Nb, Mo)

capacity is shown in Fig. 5. It will be noted that even if the region producing the largest amount of vanadium, South Africa, ceased to supply the non-communist world, there would only be a shortfall of 7%. If either of the other two regions ceased production there would be excess capacity; even on the basis of existing capacity in 1978, production outside South Africa would meet at least 76% of the consumption in that year. It is interesting to compare the position of vanadium with that of niobium and molybdenum, two metals which are often used with vanadium in alloy steels; it will be seen from Fig. 6 that the effect of an interruption in supply from the major production regions for these two alloys would be of the same order or greater than that for vanadium.

CONCLUSION
Vanadium, like all minerals on this earth, is not inexhaustible and supplies to the Western world are vulnerable to revolution, war, and international political activity; but compared with other minerals, vanadium supplies can be assured to the same degree as many minerals and better than most.

ACKNOWLEDGMENT
The author wishes to thank his colleague Mr D. Griffin for his help in obtaining the data and preparing the figures relating to production and consumption used in this paper.

P. J. Kiddle
Tungsten

As is well known, tungsten was discovered and isolated from its ore minerals, scheelite and wolframite, in the 1780s. The first significant use of tungsten was to make high-speed tungsten—manganese steel. This application was developed in the middle of the 19th century and accounted for most of the tungsten used before 1900. Demand for tungsten in armaments during World War I led to the discovery of the wolframite deposits in what is now the People's Republic of China and the subsequent development of China's

A brief historical review of the uses and development of tungsten is followed by estimates of growth rates both in the Eastern bloc and in the Western world, the latter estimated at 4·5% for the 1980s. The supply/demand picture is then clarified, including the influence of the GSA stockpile and of production in the People's Republic of China. Finally the author refers to potential new sources of tungsten, and speculates as to how these could be influenced by the economic and political climates expected up to the year 2000.

The author is managing director of Climax Molybdenum Company Ltd, London.

tungsten mining industry into the world's largest. Since the late 1920s when sintered tungsten carbide tools were established commercially in Germany, there has been a steady growth in the consumption of tungsten, especially since the end of World War II following the developments in cemented carbides which took place in the United States, Austria, Sweden, and other countries.

During the last twenty years or so, Western world consumption has grown by about 4% per year while production in the West has increased only some 3% during the same period. Future growth of virgin tungsten consumption has been forecast at only 3·2% per year between 1974 and the year 2000 in a recent study by the US Bureau of Mines which also took into account growth in the Eastern bloc countries. Another study made in the Federal Republic of Germany projected a total growth of 3·4% per year until 1990 with a rate of 4·2% for the Western world. Our forecast for the latter through the 80s is at a 4·5% rate, taking into account generally slower forecasts for economic growth in the West, greater utilization of scrap by the cemented carbide industry because of new technology and higher prices, increased usage of coated carbides for longer life, and the end of the transition from tungsten-grade to molybdenum-grade tool steels.

Because tungsten is an important strategic material necessary for defence purposes (it is estimated that 5–10% of Western consumption is for armaments), the US Government significantly expanded mine production in the USA and stockpiled tungsten in the 1950s when the supply from the People's Republic of China was disrupted during the Korean conflict. Termination of the US Government tungsten stockpile purchases in 1959 was followed by closure of most of the 750 mines in the United States and many mines elsewhere in the market economy countries. A number of these mines remain a potential source of future supply.

Since 1965, the General Services Administration (GSA) has conducted an active sales programme of excess government stocks of tungsten ores and concentrates. Most of these sales have been moderate since 1970; however, during a nine-month period between mid-1969 and early 1970 the GSA sold about 24 000 tons of tungsten during one of their disposals of stockpile supplies. At that time the GSA purchases overfilled the gap between supply and demand. However, by late 1970 the combination of deteriorating economic conditions and over-speculation on GSA material led to a rapid accumulation of inventories. The US Government's strategic stockpile of tungsten has been and continues to be an important factor in the tungsten market. While the GSA stockpile level has decreased substantially since the early 1960s from about 100 000 to about 45 000 tons at the end of last year, the surplus available for disposal, which is about 25 000 tons, currently equals about one year of Western consumption. However, it is estimated that if present US Government objectives are maintained and that if sales of tungsten continue at the recent rate of 2000–2500 tons per year, then the excess will be wiped out in the early '90s.

Let us now look at some of the vital statistics in the tungsten market today: in 1977, estimated Western world production was approximately 22 000 tons of tungsten and reached 25 000 tons in 1978. Consumption in the West was about 21 000 tons in 1977 and 23 000 tons in 1978. However, not taken into account in these numbers are about 2 000 tons of tungsten imports from the People's Republic of China in both 1977 and 1978 and also an estimated export of about 4 300 tons to Eastern Europe in both years. We also have to reckon with about 2 000 tons sold by the GSA in 1977 and 2 500 tons in 1978.

It is estimated that the United States and Canada together produced 5 500 tons, with Western Europe as a whole producing 4 000 tons and Japan about 700 tons. The remaining 15 000 tons were produced in Third World countries; in decreasing order of output, these were: Bolivia, Australia, South Korea, Thailand, Brazil, Peru, and ten or so other small producing countries.

While the major industrial countries of the West contribute only 40% of the total tungsten production, these countries account for 95% of the estimated total Western world consumption. The United States produces about 35–40% of its requirements, Western Europe 30–35% of theirs, and Japan only 15–20%.

Turning to the Eastern bloc countries, we estimate that the main producers are the People's Republic of China with about 9 000 tons, the USSR with 8 000 tons, and North Korea with about 2 300 tons. As we have seen, the USSR and its satellite countries are net importers of tungsten. These import levels are thought to exceed the material which flows from China to the West. We believe that China exports a major portion of its production, perhaps 5 000 tons, to the USSR and satellites. It is strongly felt that the Soviet authorities will attempt to become more self-sufficient in tungsten supply since they must feel vulnerable so long as they rely on China and the Western world for significant supplies of this crucial metal.

It is known that China has substantial reserves (probably 50% of world tungsten reserves) and with the rising interest among Western countries in investment there, future development of China's reserves could lead to higher production levels. However, China's goal of industrialization could lead to substantially greater levels of domestic consumption of tungsten.

Table 1   Estimated balances of tungsten supply in Western world 1977 and 1978 (tons)

	1977		1978	
Production	22 000		25 000	
From GSA stockpile	2 000		2 500	
Consumption		21 000		23 000
Exports		4 300		4 300
	24 000	25 300	27 500	27 300
Requirement from China	1 300		—	

In 1976, the Stanford Research Institute published a study entitled 'World Minerals Availability'. In the first volume they covered the economic, political, and strategic considerations for the world mineral industry, and the power politics of both the so-called 'East–West' struggle involving the USA, the USSR, and China and the 'North–South' conflict between the industrialized and less developed countries.

The Institute predicted that international politics will continue to be dominated by relations between the world's two super-powers and among the superpowers and the People's Republic of China. At the same time Western Europe, Japan, and the oil states of the Middle East will exert considerable economic influence in a world of growing interdependence, even though, on a power scale, these three centres will not approach the

Table 2   Vulnerability of Western world to interruption of tungsten supply (1978 onwards)

Producer country	Supply to Western world, tons
USA and Canada	5 500
Western Europe	4 000
Japan	700
Other sources in descending order of importance:	
Bolivia	
Australia	
South Korea	
Thailand	
Brazil	
Peru	
10 minor suppliers	15 000
	25 200 (to meet 25 000 consumption)

Failure of supply from any one of 16 countries together providing 15 000 t would have little adverse effect.

Table 3   Estimated balance between production and consumption of tungsten for Eastern bloc (1978)

Eastern bloc country	Production, tons	Consumption, tons
USSR	8 000	10 000–15 000
Satellites		5 000–10 000
North Korea	2 300	
P.R. China	9 000	?
	19 300	15 000–25 000 (plus P.R.China)

Production/consumption roughly in balance; but note that China exports 2000 t to West, while East imports 4300 t from West. Expansion in China will require increases in production as well as consumption.

United States, the USSR, or even the PRC.

In addition to the familiar East–West confrontations of the postwar world, future years will be characterized by continuing North–South tensions. The issues at stake here are the terms of trade between countries producing raw materials and the industrialized world, and appropriate levels of foreign aid and technical assistance from the latter. In the North–South conflict, the industrialized trading countries of the West are grouped together with the Communist countries. In their efforts to modernize rapidly and close the technological gap between themselves the affluent countries, the less developed world can be expected to play East against West and avoid whenever possible any commitment to any military bloc. So long as the major Western economies remain technologically advanced, less developed countries will see advantages in maintaining close relations with them. The climate for foreign investments will not be universally favourable in less developed countries, for they will continue to insist on control over their own resources. Foreign investors may be taxed heavily and expropriations can be expected to continue on an arbitrary basis. However, the vast problems confronting Third World leadership will instill a sense of pragmatism and selectivity in local attitudes to foreign investment and the acquisition of technology.

When we analyse the foregoing data and considerations, it seems clear that while additional tonnages may be predicted in the less developed countries in the future, the major development in them will be to upgrade the tungsten raw materials and even to manufacture sophisticated products which will greatly enhance both revenue and local employment and contribute to the economic and industrial development of certain Third World countries.

For entirely sound economic and strategic reasons, and because of their appropriate medium size, major productions have started during the '70s in Western Europe such as the Anglade mine in France, the Yxsjoberg deposit in Sweden, and the vertically integrated Mittersill mine, all correctly timed during a major period of industrial expansion. The same can be said for the augmented production of existing mines in Australia and Canada. In the next decade we expect that as a result of active exploration in North America, where the largest free world reserves are known to exist, operation of large mines such as the Amax Mactung property in North-West Canada will come on stream. Other promising sources are located in Nevada and new discoveries are to be expected in the Cordillera folded mountain range, particularly in the northern sections in Canada and into Alaska.

S. R. Keown
Niobium

Niobium (Nb) is one of the newer technological metals with a current world consumption of just over 11 000 tons per year and a high growth rate. Over 80% of all the niobium produced is used as an alloying element in a wide variety of steel ranging from high-strength, low-alloy (microalloy) steels which contain only about 0·04%Nb to engineering, tool and stainless grades which may contain up to 1%Nb. The metal is also used as an alloy addition to superalloys, which are mainly nickel- and cobalt-based, containing 2% - 16%Nb,

The emergence, comparatively recently, of niobium as one of the newer technological metals is briefly reviewed and its plentiful resources are indicated. The demand for niobium has increased rapidly in the last few decades, especially for steelmaking and super-alloys. Future growth in its usage is encouraged by the available spare capacity and the assurance of ample supplies for 500 years ahead at least.

The author is in the Department of Metallurgy at Sheffield University, and is Technical Advisor to CBMM (Companhia Brasiliera de Metalurgia e Mineracao, Brazil).

and in the pure form in nuclear and space technology and in superconductors. Other special applications include cermets, ceramics and glasses.

Niobium was discovered in 1801 by Charles Hatchett, an English scientist who detected it in a mineral specimen of columbite collected in New England. Hatchett named the element columbium (Cb) in honour of the discoverer of North America. In 1844 the German chemist Heinrich Rose rediscovered the element but, thinking it to be different, named it niobium (Nb). The dichotomy of the name has persisted to this day and it is the only well-known element with two names. In 1949 the International Union of Chemistry pronounced niobium to be the official international name but the element is still called columbium in the USA although European and most other countries have accepted niobium.

It was not until the early 1930s that niobium found any technological application. It was then added to tool steels as an alternative to tungsten, and to austenitic stainless steels to prevent intergranular corrosion by stabilization of the carbon in the steels. High-temperature and engineering steels were subsequently developed but the real breakthrough came with the development of high-strength, low-alloy steels in the early 1960s. Very small amounts of niobium added to mild steel were found to promote grain refinement which simultaneously improves strength and toughness.

The emergence of niobium as an important metal, and particularly as an alloy addition, has been due not only to technological innovation in the design of alloys but also to the discovery of important new ore reserves in Canada and Brazil in the 1950s. The development of suitable extraction and refinement processes has also meant that supplies of niobium are readily available at an economical cost, with vast reserves ensuring supplies well into the 21st century.

Table 1   World consumption of Nb_2O_5, tons equivalent

Year	Amount	Year	Amount
1965	4 080	1973	11 070
1966	4 760	1974	13 830
1967	4 990	1975	11 750
1968	6 120	1976	12 970
1969	7 350	1977	12 250
1970	8 440	1978	13 610
1971	8 990	1979	16 708
1972	10 570		

OCCURRENCE AND AVAILABILITY

Niobium is relatively plentiful in the Earth's crust, averaging 0·005% which is slightly more than there is of lead and slightly less than of copper. The most important ores of niobium are columbite/tantalite $(FeMn)O(NbTa)_2O_5$ and pyrochlore $NaCaNb_2O_6F$. Before 1960, columbite was the main source of supply[1] but with the discovery of richer pyrochlore deposits and suitable extraction processes pyrochlore is now by far the most important niobium-containing mineral, currently supplying 85% of world demand. The main deposits of pyrochlore are in Quebec, Canada, mined by Niobec Inc. and in the state of Minas Gerais in Brazil mined independently by Mineracao Catalao des Goias and by Companhia Brasileira de Metalurgia e Mineracao (CBMM). At the present time deposits of pyrochlore in Zaire are not being commercially developed. Brazil produces over 70% of the world's niobium and their pyrochlore has the highest niobium content, averaging more than 2·5%Nb_2O_5.

To give an example of the future availability of niobium, CBMM, who are the largest producers — having satisfied 65% of total world demand over the last 10 years[2] — have cautiously estimated that the ore reserves at their Araxa mine are at least 460 million tons.[3] At twice the present world consumption rate this will last for over 500 years and there are other areas in Brazil with high potential for additional reserves.

The pyrochlore ore is concentrated by selective flotation and leaching to give, at the CBMM Araxa plant, pyrochlore concentrates averaging 63%Nb_2O_5. Most of the concentrate is reduced by the aluminothermic process to give ferroniobium suitable for steelmaking and containing about 66%Nb. Some high-alloy steels and all superalloys demand higher-purity master alloys of FeNb, NiNb, CrNb, etc. which are produced from high-purity niobium oxide of various qualities reduced from the pyrochlore concentrate. Niobium oxide of the very highest purity is required to produce niobium metal for nuclear applications, aerospace, and superconductors.

At the present time the CBMM plant has a capacity of 32 million lb of Nb_2O_5 per annum out of a world capacity of about 44 million lb. By 1981 the CBMM capacity will be 55 million lb per annum with further plans to increase this to 74 million lb when required.[4] The world consumption of Nb_2O_5 from 1965 to 1979 is shown in Table 1.

APPLICATIONS AND MARKET DEVELOPMENT

The usage of niobium varies significantly from country to country depending on such factors as the level of technology, specialization of products, and traditional markets. The only

99

Table 2   Niobium consumption in the USA,
Nb contained, tons

	Year			
Application	1975	1976	1977	1978
Microalloy steel	847	854	1 070	1 366
Stainless and heat-resisting steel	205	208	258	365
Full-alloy steel	188	239	265	298
Superalloys	202	191	361	520
Other alloys	37	23	30	27
Miscellaneous	40	23	9	7
totals	1 519	1 538	1 993	2 583

reliable data on the breakdown of niobium consumption are those for the USA analysed in Table 2. This shows that about 80% of all niobium is used in steelmaking and most of the remaining 20% for non-ferrous applications is used in superalloy production. Breaking down the steelmaking portion still further, it can be seen that microalloy steels represent a major share of the total niobium market. It is worth examining the remarkable growth of these steels in some detail, since their past and future development have a major effect on the demand for niobium.

Microalloy steels

Although patents for the addition of small amounts of niobium to plain carbon steel were granted in 1941 it was not until 1957 that trials conducted in Pittsburgh demonstrated that the strength of structural plate steels could be increased by additions of only 0·015%Nb.[6] Colvilles Steelworks in Motherwell pioneered the use of Nb in steels in Europe in the early 1960s[7] and 'micro' additions of about 0·1% or less of niobium, vanadium, and titanium were soon being added to a wide range of steel products such as strip, sheet, bar and plate to improve the strength and sometimes also the toughness of these essentially mild steel compositions.[8] The physical metallurgy of these alloys is quite complicated[9] and will not be described here except to say that the microalloying elements precipitation-harden and prevent grain growth during heat treatment by the formation of alloy carbides of the forms NbC, V_4C_3, and TiC. Niobium has one further unique advantage in that it will produce additional grain refinement in steels (to give increased strength and toughness) controlled-rolled to low finishing temperatures, by its delaying effect on austenite recrystallization (Fig. 1).[3]

Microalloy steels are now replacing mild steels in a wide variety of applications and it should be remembered that mild steel accounted for about 90% of all iron and steel products before the advent of microalloy steels. The specifications for products such as pipelines, reinforcing bar, structural sections, automobile steels, etc., are being continually upgraded to meet the demands of higher strength/weight ratios, energy-saving requirements and more stringent safety regulations. The addition of 0·03%Nb adds only £6 to the cost of one ton of steel, so that microalloying is the most economical method of meeting the improved specifications when compared with conventional alloying or subsequent heat treatments.

The consumption of niobium can be seen in Fig. 2[3] to be related to the production of raw steel and the future growth of niobium should therefore be related to the growth of steelmaking.[3] Assuming that the developing countries adopt microalloying technology (and there is clear evidence that this is occurring) the current steelmaking recession

CONTROLLED ROLLING   (a) carbon–manganese steel

immediate

reheating → rolling → recrystallization → transformation
austenite   austenite   austenite   ferrite

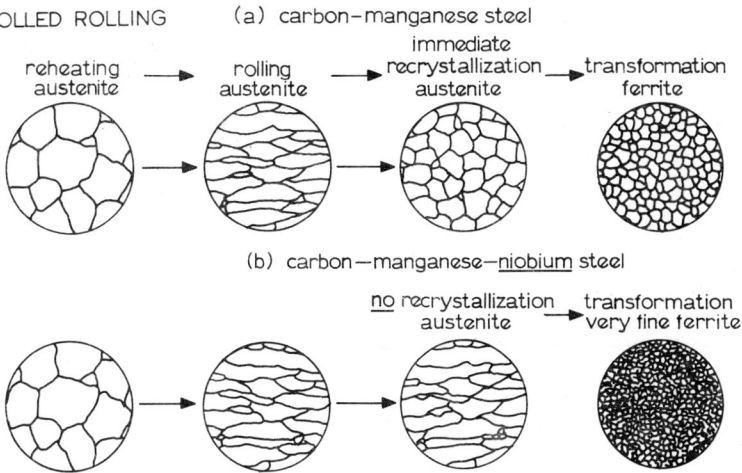

(b) carbon–manganese–niobium steel

no recrystallization   transformation
austenite   →   very fine ferrite

1   **Influence of niobium on grain refinement of controlled-rolled mild steel (Ref. 3)**

in the Western world should have little effect on microalloy consumption. Additionally, more stringent specifications for components and recent trends in the USA and in Japan to double or triple the niobium content of certain steels could easily increase the average growth rate of about 8% experienced since 1967. In fact the current rate of growth is much higher than 8% (Table 1).

Other alloy steels

There has been some reduction in demand for niobium in stainless steel during the 1970s due to the introduction and adoption of the AOD steelmaking process by nearly all the world's stainless steelmakers. This process allows the carbon level of stainless steels to be reduced to very low levels so that stabilization of the carbon by niobium or titanium is no longer required for certain applications.[10] However, niobium is still very important for creep resistance and for acid corrosion resistance in some austenitic stainless steels and for greater high-temperature strengthening in martensitic stainless steels. The emergence of ferritic stainless steels in the last few years

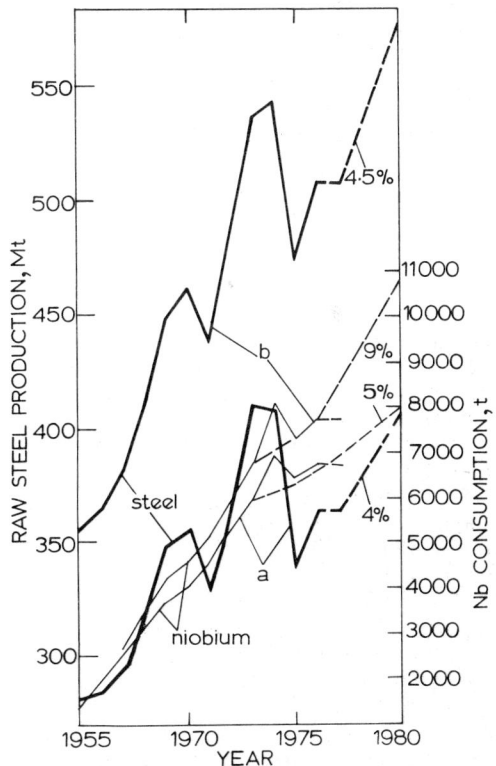

2   **Correlation between raw steel production and niobium consumption (Ref. 3)**
(a) W. Europe, N. America, Japan
(b) World excluding China, Russia, N. Korea

101

is also likely to provide an important market for niobium.[11]

In low-alloy engineering and creep-resisting grades and in some low-alloy tool steels there is a trend towards the substitution of expensive alloying elements and an increasing awareness of grain control by microalloying technology. Many of the low-alloy grades were developed before niobium was readily available as an alloying element and it is now becoming realised that niobium, in addition to being relatively inexpensive and a strong carbide-forming element which increases strength, wear resistance and high-temperature strength, can have a significant effect on the hardenability of steels.[12]

One of the potentially most important new markets for niobium is in the substitution of relatively large amounts of tungsten, molybdenum, and vanadium by the less expensive niobium in high-speed tool steels.[13] Current trials are indicating that niobium can be used to replace any of these elements in M2 high-speed steels with very significant cost saving benefits.

Other applications

Although alloy steels are the most important market for niobium a significant tonnage of niobium is used for non-ferrous applications. Of this, by far the greatest amount is used in superalloys, about half of this being utilized in jet engine components. This market is expected to increase significantly as is the use of superalloys in traditional and nuclear power generating plant, raising the demand for high-purity niobium oxide. Only a few tons of high-purity niobium metal is used in the more specialized scientific applications such as superconductors, nuclear fuel canning, and space exploration activities and no great expansion is envisaged. Just recently the high cost of tantalum has encouraged the hard-metal tools industry to consider the substitution of tantalum by niobium carbide in cermets. Trials are also going ahead to consider niobium for coinage applications. Tests carried out at the Philadelphia Mint about 15 years ago clearly established that niobium has excellent coinability properties in terms of weight, lustre, ring, hardness, and difficulty in counterfeiting. Shortages and price escalations in traditional coinage materials could lead to niobium being used for high-denomination coins, so greatly increasing the demand for niobium.

CONCLUSIONS

In the last twenty years niobium has emerged from being an expensive, scarce, scientifically interesting but technologically little-used metal[14] into its present status as one of the most important alloying elements in steels and superalloys. This has been brought about by the discovery of huge new reserves of pyrochlore in Brazil and Canada and by the successful development of economical, large-scale production processes to ensure plentiful supplies of niobium oxide, ferroniobium, and niobium metal well into the next century. Technological innovation and the development of new alloys with improved properties has been equally important in that this has increased the demand for niobium, providing an incentive to niobium producers to expand their activities.

It seems probable that on-going metallurgical developments — and here one should stress the importance of metallurgical research and industrial collaboration — and problems in the supply and cost of other metals will increase the demand for niobium. It should be emphasized that there has never been any supply problem with niobium and that ambitious current expansion plans by the producers should ensure that this situation continues. It is also important to appreciate that the price of niobium has always been stable, with only gradual price increases and none of the erratic fluctuations encountered with many other metals at the present time.

In view of the anticipated demand for niobium and the availability of sufficient spare capacity to meet any sudden increase in demand, the niobium industry is looking ahead with confidence, having more than adequate supplies for at least 500 years.

REFERENCES

1 G. L. MILLER: 'Tantalum and Niobium', 1959, London, Butterworths
2 *American Metals Mart*, 24 August 1978, 12
3 F. HEISTERKAMP AND H. STUART: Proc. European Conference on Ferroalloys, 1977, Zurich, Switzerland
4 CBMM Press Release, *American Metals Mart*, 29 June 1979, 2
5 United States Bureau of Mines Yearbooks 1975–1978
6 W. G. WILSON: private communication
7 W. B. MORRISON: *J. Iron Steel Inst.*, 1963, 201, 317
8 (a) ISI Special Report No. 81, 1963, 'Metallurgical developments in carbon steels'
(b) ISI Special Report No. 104, 1967, 'Strong, tough structural steels'
(c) Proc. Metallurg. Companies Conf. November 1970, 'Low-alloy high-strength steels'
9 S. R. KEOWN: Proc. 3rd CBECIMAT Congress, 1978, Rio de Janeiro, 2, 1
10 M. A. STREICHER: Proc. Conf. 'Stainless Steel '77', 1977, Climax, 1
11 R. M. DAVISON AND R. F. STEIGERWALD: *Metal Progress*, 1979, June, 40
12 J. H. WOODHEAD: private communication
13 F. HEISTERKAMP AND S. R. KEOWN: *Metals and Materials*, October 1978, 35
14 B. B. ARGENT: *J. Inst. Met.*, 1957, 85, 547

J. Nutting
Cobalt and chromium

At first sight cobalt and chromium may appear to have little in common but, as transition metals, they are separated only by iron and manganese in the periodic table. Neither metal is widely used on its own and they are not particularly good in providing a base for alloys, although a cobalt–chromium alloy is widely used in dentistry and for surgical implants. The chief uses of the two metals are as alloying elements with a variety of bases

---

The international significance of cobalt and chromium is discussed in relation to the political uncertainties which surround the supplies of both metals. The resources of cobalt seem assured far into the future and while recent price levels have restricted some usages, the demand is not price-sensitive in applications such as high-temperature alloys in gas turbines. The availability of chromium is highly dependent on political consider-ations since almost all the known reserves are in Southern Africa and the Western world has to import virtually all the chromium it needs.

Professor Nutting is Head of the Department of Metallurgy, Houldsworth School of Applied Science, University of Leeds.

and, in many of the materials so produced, substitution would be difficult and costly. Thus it is not surprising that there is much interest in the availability of supplies, their source, and the medium- to long-term price changes.

With both metals difficulties arise from the distribution of the major ore deposits. In the case of cobalt there is a close link with the copper deposits of Central Africa, and politi-cal actions in this region have had a marked influence on availability, and hence on world prices. With chromium the major sources are Zimbabwe-Rhodesia and South Africa and, with politically inspired sanctions in the former and political uncertainties in the latter, it is to be expected that the major industrial groups in the USA, the EEC, and Japan, the chief users of chromium and all with a heavy dependence on imports, should express concern about future supplies.

COBALT

Cobalt has a relatively long resource life, as can be seen from the data given in Table 1, which is derived for the land-based minerals. However, cobalt is also present in significant amounts in the manganese nodules found in the deep ocean beds and, although these

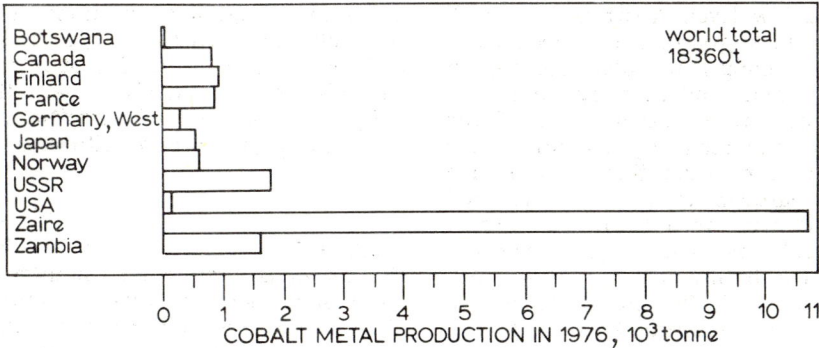

Botswana
Canada
Finland
France
Germany, West
Japan
Norway
USSR
USA
Zaire
Zambia

world total
18360t

0 1 2 3 4 5 6 7 8 9 10 11
COBALT METAL PRODUCTION IN 1976, 10^3 tonne

1   Sources of cobalt production 1976

nodules are not harvested at the moment, there seems to be little doubt that they will be once the legal problems of ownership are overcome. It is postulated by some authorities that manganese nodules are being continuously generated at the root of the deep ocean trenches by a volcanic-type activity; therefore the long-term supply of the metal seems assured.

A broad classification of the uses of cobalt is given in Table 2. It has been argued that of these uses about one-third are critical; that is to say, substitution would be difficult or costly. The majority of the critical uses relate to the addition of cobalt to the high-temperature nickel-base alloys used for blades and discs in both aircraft and land-based gas turbines. As there is a high added value to the basic materials cost of a turbine, the demand for cobalt for this application is not price-sensitive and therefore, in times of cobalt shortage, it is the engine builders who dictate the price since they try to ensure that they are not short of this vital metal.

In 1976 world production of cobalt was 18 360 tonnes. The chief sources of cobalt are given in Fig. 1 which shows the important supply role played by the Central African countries of Zaire and Zambia. It is not surprising, therefore, that when cobalt production in Zaire was stopped a short while ago there was an immediate shortage of cobalt. The position was aggravated by the fact that cobalt production from other sources was also

Table 1   Static resource life (in years) for some widely used metals

$10-10^2$ years	10^2-10^3 years	10^3-10^4 years	$>10^4$ years
W	Cu	Cr	Co
Mo	Ag	Al	Fe
Au	Sn	Ni	Mn
Pb	Zn	Ti	

Table 2   The uses of cobalt

Use of cobalt	%
Magnets	26
Superalloys	20
Cemented carbides	15
Tool steels	5
Catalysts	7
Pigments	7
Metal-organics	20

running at a low level. Apart from a small direct source in Morocco, cobalt occurs as a by-product of copper manufacture, e.g. in Zaire and Zambia, and as a by-product of nickel manufacture in other parts of the world. During the last few years with a poor world market for nickel there has been a consequent curtailment in the associated cobalt production. Accordingly the market price for cobalt has risen sharply. For many years there has been a fairly constant price ratio of about 2:1 for cobalt to nickel. Recently this ratio has risen to 10:1 but the signs are that, with the return to more normal working conditions in Zaire and the recent increases in the price of nickel, the availability of cobalt will improve and the Co:Ni price ratio will fall. However, it seems unlikely that the old ratio of 2:1 will be established in the near future.

The airlines of the world are faced with aging fleets which will shortly have to be replaced. This will lead to a demand for engines and hence for cobalt. But there are still so many uncertainties in Central Africa that the supply situation is not assured, hence prices could fluctuate.

The price levels already established have had a strongly negative influence on the demand for cobalt for polymer additives and pigments; that is to say, for dispersive applications. On the other hand, in many of the major metallurgical applications of cobalt it is possible to recover and recycle the metal. A real price increases the incentive to recycle and this can be no disadvantage to the availability of cobalt even bearing in mind the excellent long-term supply situation.

CHROMIUM

Chromium is vital to our industrial society and if supplies are curtailed then the immediate consequences could be severe. In a recent analysis prepared in West Germany it was suggested that if their chromium supply were cut by 25% there would be an immediate decrease in their gross national product of about 15%. The long-term supply situation in relation to the reserves is not too difficult, as can be seen from Table 1. The real problem arises from the fact that 96% of the known reserves are in Zimbabwe-Rhodesia and South Africa and, with political uncertainties still far from resolved in these countries, deep concern is expressed about future chromium supply.

In 1976 the annual world production of chromium was 8·4 million tonnes. The chief producers are listed in Fig. 2, from which it can be seen that the USA and Western Europe are almost completely dependent upon imported chromium.

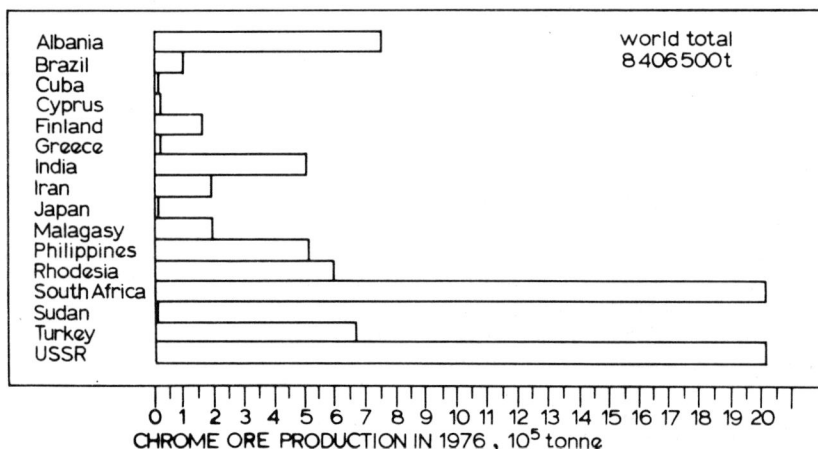

2   Sources of chromium production 1976

Table 3   Chief uses of chromium

Use of chromium	%
Metallurgical chromium, ferroalloys, charge chrome	60
Refractories	20
Chemical industry	12
Foundry sands	8

The chief uses of chromium are given in Table 3, which exhibits the important role of chromium as an alloying addition. A considerable proportion of chemical chromium is also used in the metallic state, via electrolytic and aluminothermic chromium for addition to high-temperature alloys. Metal plating and surface treatment solutions account for about a quarter of the chemical uses of chromium. The remainder of the uses are for pigments and for the tanning of leather.

The use of chromium as a chromite refractory has decreased with the demise of the basic open-hearth steelmaking furnace. However, chromite is still used as chrome-magnesite in the walls of electric arc furnaces. If chromite became scarce then no doubt other refractories could be developed to replace it; but even if more widely available, they are likely to be more expensive. Chromite as an additive to foundry sands has developed in competition with the more expensive zircon, but there is now a glut of this latter material arising from the improved separation processes developed for illmenite-bearing beach sands. Thus substitution of chromite for this application would not be too difficult.

There are three major metallurgical roles of chromium. In solution in austenite it slows down the $\gamma \rightarrow \alpha$ reaction, so increasing the hardenability of a steel. Many other common alloys exert a similar effect and therefore as an alloying addition for hardenability chromium could easily be dispensed with. Chromium is a stronger carbide former than iron and the carbides produced such as Cr_7C_3 and $Cr_{23}C_6$ are harder than Fe_3C. Thus improved hardness and wear resistance is obtained by the addition of chromium to steels used in the quenched and tempered condition. There are many other elements which are even stronger carbide formers than chromium and which give carbides which are harder than chromium carbides. Thus, if chromium supplies were curtailed, alternative but more costly steels could be developed for applications such as ball bearings and ball races. The third and perhaps most significant use of chromium is as an alloy addition to give resistance to corrosion and high-temperature oxidation, particularly to steels but also to nickel-base alloys.

It is perhaps ironic that the benefit conveyed by chromium in these applications depends entirely on the low ionic transport properties of Cr_2O_3 present as a thin film on the surface of the alloy yet, in a standard stainless iron or steel, 12–18% chromium is required as an additive to achieve a suitable film. Other metals have stable oxides which could confer oxidation resistance, but unfortunately their atomic diameters are different from that of iron and hence their solubility in iron is lower and their ability to form new phases is higher than that found with chromium which has an almost identical atomic diameter to that of α-iron. Thus, if there were insufficient chromium to meet the demand for corrosion-resistant and high-temperature steels, substitution would have to be via entirely different and more expensive (or less satisfactory) alloys such as those based on copper, aluminium, or titanium.

In some respects there has been an alleviation of these problems by the development of the argon oxygen decarburization (AOD) process because this enables lower-quality chromium-containing materials with a high carbon content to be used in the manufacture of stainless steel.

107

But what of the long-term future? Agricola, in his famous treatise 'De re metallica', made the profound observation '. . . that where one ore is found another is not far away.' In the light of modern theories of continental drift, this statement has to be viewed from a rather different standpoint than in Agricola's day, and it is not surprising that geologists are now looking for the ultramafic rocks associated with chromite deposits in Western and Central Australia as a continuation of the mineralized region of the Bushveld complex of South Africa and the Great Dyke of Rhodesia.

The severe price rises of chrome ore which occurred in 1975 have not been maintained, partly as a result of the general recession in world trade; but the immediate future depends more on political settlements than conventional economic arguments and therefore the non-communist world must endeavour to maintain access to African supplies of chromium for a long time to come if it is to maintain the standard and type of life it has become used to enjoying.

SOURCES

J. NUTTING: *Metals and Materials*, 1977, July/August, 30

E. T. HAYES: 'Superalloys: metallurgy and manufacture,' Proc. Third Int. Symposium, September 1976; 1976, Louisiana, USA (Claitor's Publishing Division)

J. EDWARDS AND P. ROBBINS: 'Guide to nonferrous metals and their markets'; 1979, London, Kogans Page

W. OMER COOPER AND P. G. MOESKOPS: Amdel Bulletin, Australia, 1978 No. 23

M. T. Herbert,
G. J. K. Acres, and J. E. Hughes
Platinum, gold, silver, and mercury

While mercury has little in common with the precious metals, in one respect, namely recovery and recycle, important similarities exist. An essential feature of the use of the precious metals in most industrial applications is the recovery of the spent material to be recycled and processed for further use. In most industrial applications the use of the precious metals is only economically viable if recovery and recycle is practical. As far as conservation of the strategically important materials is concerned, recovery and recycle is essential to ensure their continuing availability.

The estimated reserves (currently mined deposits) and resources (unmined deposits) of these elements, together with a description of their main applications is given in the sections which follow.

PLATINUM AND
PLATINUM GROUP METALS

Platinum and the other five platinum group metals (PGMs) — *palladium, rhodium, iridium, ruthenium, osmium* — commonly occur together in nature and are among the scarcest of metallic elements. In the richest deposits their total concentration rarely exceeds 10 ppm. The reserves and known resources of these elements occur in South Africa, the USSR, and Canada which together currently produce 99% of the world's newly mined PGMs. In 1978 the total output from these countries was 198 700 kg, of which 46% was mined in South Africa, 46% in the USSR, and 6·5% in Canada. The remaining 1·5% was obtained from other sources.

In South Africa the PGMs are currently mined from the Merensky Reef as principal

---

In the context of future availability, the reserves and resources of the precious metals — platinum, silver, and gold — are reviewed geographically. The current industrial applications of these elements are described as a background to their future availability. In addition, new applications for the precious metals are reviewed. No foreseeable shortfall in the availability of the precious metals is forecast. In addition to the precious metals,

the supply/demand situation for mercury is reviewed. The previously predicted shortfall in the availability of mercury is discussed by reference to changes in its industrial applications resulting from the introduction of new technology.

The authors are with Johnson Matthey and Co. Ltd, Research Centre, Reading, UK.

Table 1 (South Africa)

Metal	%
Pt	60
Pd	25
Rh	4
Ru, Ir, Os	11

Table 2 (USSR)

Metal	%
Pd	56
Pt	30
Rh	10
Ru, Ir, Os	4

products, with copper, nickel and cobalt as co-products. The Merensky Reef, which was first mined in 1924, is estimated to contain most of the world's reserves and one-fifth of the world's resources of PGMs. The Reef occurs in the Bushveld igneous complex, an oval-shaped basin of igneous crystalline volcanic rocks covering an area of 100 × 200 miles. Current PGM production from this source is approximately 93 000 kg/annum, the composition of which is shown in Table 1. As a result of variations in the proportion of the PGMs in the different deposits, while South African total production of PGMs represents 46% of the current world output, South African output of platinum represents 65% of total world output.

In the USSR currently the major source of PGMs is in the nickel-mining complex in North Western Siberia. The USSR produces annually a similar amount of platinum group metals to that produced in South Africa, namely 93 000 kg/annum which is principally palladium with platinum and rhodium as relatively minor constituents. The composition of the output from USSR sources is shown in Table 2.

Canadian production of PGMs is a by-product of nickel mining where, in the richest mines, approximately 75 g of PGMs are produced for every ton of nickel extracted. Total production of PGMs is small compared to that of South Africa and the USSR and output is determined by nickel operations, unlike the South African activity where the ore is mined primarily for its PGM content.

Estimates of world reserves and resources of the platinum group metals appear in Table 3. US resources, which are sizeable, are concentrated in Montana, Alaska, and Minnesota. Of these resources only about 31 tonnes can be called reserves. Most of the reserves consist of by-product platinum group metals in copper reserves in the Western States; a smaller amount is in the placer deposits at Goodnews Bay, Alaska. The estimated resources of Canadian and USSR platinum group metals are virtually all by-products of nickel mining; the resources listed in Table 3 are based mainly on estimates of nickel resources. Canadian reserves are mainly at Sudbury,

Table 3 World PGM resources (tonnes)

	Reserves				Other platinum group resources	Total platinum group resources
	Platinum	Palladium	Rhodium	Platinum group		
North America:						
United States	n.a.	n.a.	n.a.	31	9283	9300
Canada	124	124	*	280	217	500
South America:						
Colombia	*	—	—	*	124	155
Asia: USSR	1860	3725	124	6200	6200	12420
Africa:						
Rep. South Africa	10860	4660	620	18000	43470	62095
Rhodesia	n.a.	n.a.	n.a.	n.a.	3100	3100
World total (rounded)	12900	3510	740	24500	62400	87570

n.a. = not available; asterisk (*) = less than 30 units

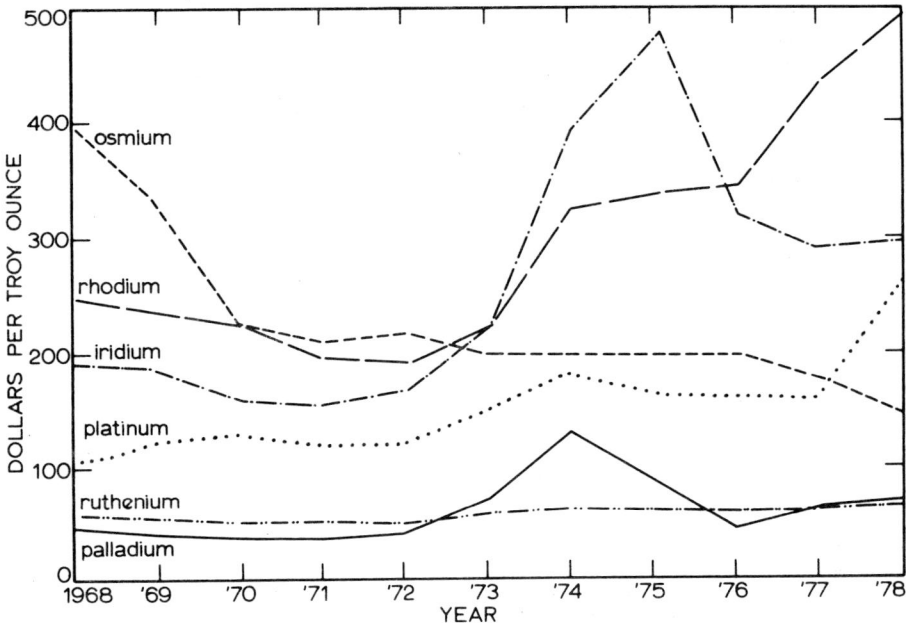

1   Average price of PGMs 1968–78

Ontario, and in the Lynn Lake–Moak Lake region of northwestern Manitoba. USSR reserves are located mainly in the Norilsk–Talnakh nickel area of northwestern Siberia, with lesser tonnages at Pechenga and Monchergorsk in the Kola Peninsula.

The huge resources of the Republic of South Africa are in three horizons in the Bushveld igneous complex, in Transvaal Province. The first horizon, known as the Merensky Reef, has been mined for decades and is extensively mapped. The other horizons, the 'Upper Group chrome seams' and the 'Platreef' have been little explored and are unmined. Based on a mining depth of 1 200 m, the potential resources of the three horizons are: Merensky Reef 18 000 tonnes; Upper Group 31 000 tonnes; Platreef 12 500 tonnes. In Table 3 the ore in the Merensky Reef is considered to be South African reserves, while the other two ores are considered to be resources. From these data it may be seen that the world's major reserves and resources of the PGMs occur in South Africa. Although very little of the current output of PGMs

arises in America a significant resource exists in that area. Taken together at current levels of PGM output known reserves would last for an estimated 150 years while total resources, assuming that the metals can be extracted economically, would last for 400 years.

In real terms the price of platinum has been relatively stable over the past twenty years while those of palladium and rhodium increased sharply in 1974. The price of palladium subsequently declined but that of rhodium has remained at the higher level, as shown in Fig. 1. The prevailing price of the PGMs together with the elements with which they are associated in the major deposits have influenced, and are likely to continue to influence, the supply/demand situation.

Current applications of PGMs
The uses of PGMs in industry are related to their catalytic activity, chemical inertness over wide temperature ranges, and high melting points. In some applications it is the

111

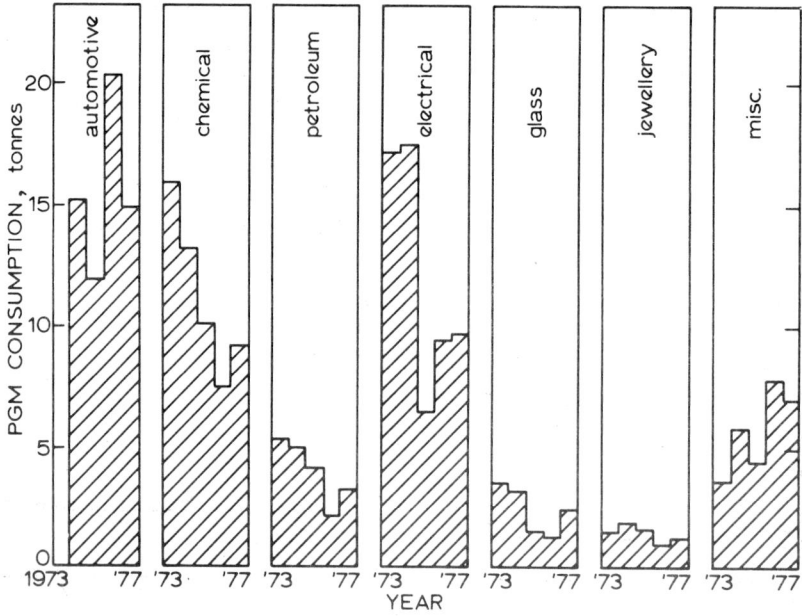

2   **Consumption of PGMs in USA 1973–77**

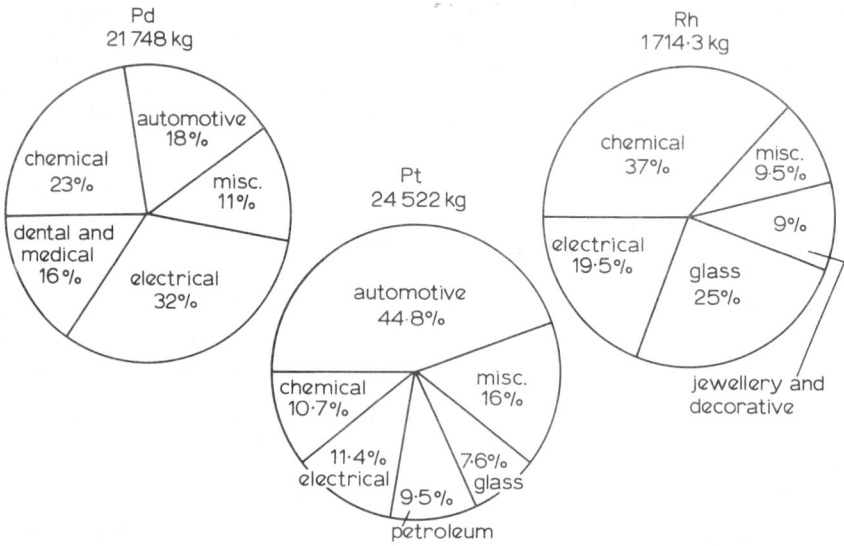

3   **Use of Pd, Pt, and Rh in 1977 (USA)**

combination of two or more of these characteristics rather than any one alone which make the PGMs uniquely useful and of strategic importance in industry as it presently exists or is likely to develop by the year 2000. Figure 2 shows the consumption of PGMs in the US by industry from 1973 to 1977 while Fig. 3 highlights the major applications of platinum, palladium, and rhodium.

Chemical industry

In the chemical industry PGMs are used as catalysts in the manufacture of heavy and fine chemicals and pharmaceuticals. The manufacture of nitric acid by the oxidation of ammonia over platinum/rhodium gauzes is an important catalytic use. Platinum catalysts are also used in the manufacture of intermediates in the plastics, synthetic rubber, and pesticides industries. Platinum and its alloys have the greatest corrosion resistance of any readily workable metal. As such they are employed extensively in laboratory and process equipment for the chemical industry. The 1977 demand by the US chemical industry totalled 8 755 kg.

As a result of energy and environmental considerations, the application of PGMs particularly as catalysts is currently increasing and the trend is expected to continue. Of particular relevance in this area are the PGMs other than platinum. For example, recent developments in homogeneous catalysis are centred mainly on rhodium which possesses unique catalyst properties presently unobtainable with any other element or combinations thereof. Based on known reserves/resources and taking into consideration the high recovery rate of PGMs used in the chemical industry, no foreseeable shortfall in supply is predicted.

Petroleum

In the late 1950s the demand for platinum was substantially increased by the development of a new catalytic process to upgrade the octane rating of petrol. This constituted the first major use of platinum in the oil industry and has subsequently contributed to an overall increase in the efficiency of the spark ignition engine. Since the introduction of the first platinum catalyst, improved systems have been developed such as platinum/iridium and platinum/rhenium which on account of greatly extended life have reduced the consumption of platinum in this application.

As a result of environmental considerations the petroleum industry is moving towards low-lead and lead-free fuels which would leave platinum-catalysed reforming as the principal method for increasing octane ratings in the absence of lead additives. This would increase the future demand for platinum. Future demand will be lower, however, if lead compounds are retained, if engines capable of using low-octane fuels are developed, or if diesel engines capture a larger proportion of the market. The use of alcohols as octane improvers could also influence the use of PGMs in this application. In practice it is likely that a combination of these options will emerge so as to maximize energy conservation in the transport area and overall it is expected that the use of PGMs will increase.

Electrical and electronics

Chemical inertness and thermal stability justify the use of PGMs in a wide range of electrical and electronic devices, e.g. relays, thermostats, thermocouples, printed circuit boards, electron tubes, resistors, spark plugs, and high-power magnets. The creep resistance and high work functions of platinum make it suitable for use in electronic grid devices and its strong resistance to electrical and chemical erosion make it suitable for aircraft and high-performance spark plugs. In low-voltage, low-current relays which are required to work reliably for extended periods, PGMs (usually palladium) are widely used. In recent years palladium has been used in place of gold owing to changes in the relative prices of the two metals. Telephone relays account for a major part of the palladium used in contact applications. In higher-voltage relays, the hardness and high melting points of platinum/iridium and platinum/ruthenium alloys minimize wear and arc transfer of metals between contacts. PGMs are ideal for use in thermometry where platinum/rhodium thermocouples are used in many applications, particularly those concerned with temperature measurement above 1000°C. Platinum is also widely used in resistance thermometers.

The future use of PGMs in electrical devices will continue at the present levels if the economy continues to expand and the use of domestic appliances and similar products continues. Demand will be decreased by the use of solid-state switching systems instead of electromechanical contacts. This can already be seen in the telecommunications industry where new telephone exchanges now employ solid-state devices. Some compensation of this trend is likely to occur with the widespread application of microprocessor technology. It is unlikely that any major new uses for the PGMs will develop in this area but it is important to note that unlike other applications for the PGMs most of the metal used in electronic systems will not be recovered and recycled.

Glass

The most important use of PGMs in the glass industry is in glass fibre manufacture, where platinum/rhodium 'bushings' are resistance-heated to 1500°C and used in the process of fibre drawing. PGMs are also used in crucibles and glass melting furnaces for glasses which must not be contaminated or discoloured, such as optical glasses. The glass and ceramics industries also decorate china and glass tableware with PGMs. Demand for PGMs in the glass industry is likely to increase as glass fibre continues to replace other textile fibres, if glass reinforced plastics replace structural metals, and if high energy costs lead to a greater use of glass fibre insulation. This constitutes a further example of the increased use of platinum group metals aimed at energy conservation. It is predicted that current energy sources will reach a shortfall position long before this situation arises with the platinum group metals. In the glass industry a high proportion of the platinum group metals in use are recovered and recycled.

Jewellery

Platinum is not at present extensively used in jewellery outside of Japan. Japanese consumption is, however, a major outlet for platinum, accounting for some 30% of total world output. Use of the other PGMs in this area is not so significant.

Future consumption of platinum by the jewellery industry worldwide is expected to increase as a result of the marketing activities of the South African platinum producers and the introduction of hall-marking for platinum in 1975. While in the short term this is considered by some to be an extravagant use of strategic materials, in the long term (as with gold and silver) this constitutes a further source of PGMs for future industrial applications.

Automotive emission control

Severe atmospheric pollution problems in the 1960s and early '70s in several parts of the world, but notably in the first instance in Los Angeles, caused both the American and Japanese Governments to introduce legislation to control exhaust emissions from motor vehicles. To reduce the levels of carbon monoxide and unburnt hydrocarbons the system which was adopted was to oxidize them to carbon dioxide and water over a platinum or platinum/palladium catalytic converter:

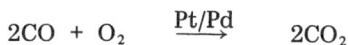

$$2CO + O_2 \xrightarrow{Pt/Pd} 2CO_2$$

and

$$HC + O_2 \xrightarrow{Pt/Pd} CO_2 + H_2O$$

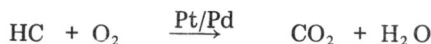

The first catalytic converters were introduced in the US and Japan in 1974. They are currently fitted to over 90% of all cars sold in America and to a substantial proportion of those sold in Japan. Currently catalytic converters are the largest single consumer of PGMs in the US. In 1977 this application accounted for 32% of total PGM usage.

More stringent controls on exhaust emissions are being introduced in America which require further reductions in CO and

HC levels, as well as a sharp reduction in NO_x emissions. The US Clean Air Act Amendments Bill of 1977 laid down the following exhaust emission standards for new cars to be sold in America over the next four years:

Model year	HC	CO	NO_x
1978/79	1·5	15	2
1980	0·41	7	2
1981 +	0·41	3·4	1

These figures refer to the permissible level of emissions in g/mile. In order to meet these new requirements three-way catalysts which contain rhodium in addition to platinum have been developed. As a result of this, the amount of PGMs currently used in catalytic converters is likely to increase as the three-way catalyst system comes into general use in 1980–81. Presently the ratio of rhodium to platinum used in some catalyst systems exceeds that of the ore from which the metals are extracted. Recent developments in catalyst technology now make it possible for the emission standards to be met without exceeding the rhodium/platinum mine ratio. This example illustrates the importance in PGM utilization of maintaining the major applications for the PGMs in balance with the proportion of the elements which occur in the major deposits as listed in Tables 1 and 2.

In the US and Japan the use of PGM catalyst systems for emission control is established. It is currently the most cost-effective system in terms of both emission control and energy conservation, and is projected to continue in use into the late 1980s. In Europe PGM catalyst systems may be adopted for emission control in the mid-80s onwards as legislation follows that in the US and Japan. As this application presently represents the biggest single use of the PGMs, consideration is being given to their recovery and recycle. While technically this is possible, the problems of collection together with economic viability remain to be surmounted for this to become a commercial operation. However, as the car industries are now incorporating into the design of motor vehicles concepts for the easy recovery and recycling of strategically important materials it is anticipated that the PGMs used for emission control will be recovered and recycled.

Fuel cells

The concept of direct conversion of fuel into electricity has been known for many years but to date exploited on only a limited scale. The advantages of the fuel cell which employs catalytic electrodes to convert fuel and air directly into electrical power are energy conservation and environmental control. While many experimental fuel cells have been built the first practical applications were as power sources in space exploration vehicles and military applications. United Technologies Corporation are now constructing a 4·8 MW fuel cell in New York for evaluation by Consolidated Edison as a peak-lopping station. The amount of platinum used in the catalytic electrodes of the cell has now been reduced to a point where the overall economics of power generation by this means are cost-effective. Such systems, being currently some 30% more efficient in energy consumption, are projected to play a major part in America for energy conservation. Coupled with their total environmental acceptance a major new application for PGMs may arise in the mid-80s. At current electrode loadings, which are reported to be of the order of 10 kg of platinum per megawatt of installed capacity, with the possibility of fifty 27 MW fuel cell power plants being built in 1985 and up to 1460 in 1990, this represents an annual requirement of between 786 and 58 690 kg of platinum per year.

This provides a particular example of the use of one strategic material, platinum, to conserve another, namely conventional carbonaceous fuels, while at the same time having a totally acceptable environmental process. It is expected that no significant loss of platinum will occur in a full-scale system and the metal will be recovered and recycled.

Miscellaneous uses

The PGMs have many uses in dentistry and medicine. In dentistry they are used as additions to gold-based alloys, in orthodontic devices, and in dental porcelains. In medicine they are used in cautery points, hypodermic needles and cardiac pacemakers. *Cis*-dichlorodiammineplatinum (trademarks 'Cisplatin and 'Neoplatin') has recently been introduced as

115

an anti-cancer drug, and although potentially important medically the amount of platinum used will be negligible compared to total consumption. Other uses for the PGMs include spinnerets for synthetic fibres, gas detection systems, bursting discs, brazing alloys for jet engines, portable hydrogen generators, and galvanic systems for protecting ships and pipelines against corrosion.

Secondary sources and recycling of PGMs
The high cost of the PGMs ensures that every effort is made for their efficient use, conservation and recovery. This task is made easier by the fact that in most uses PGMs are virtually indestructible and non-dissipative. Also, the relative ease with which they can be separated from less chemically stable materials allows almost complete recovery in most applications.

Future supply of PGMs
At the present rate of platinum production, taking into account possible major new applications, there are known reserves and resources to last at least 150 years. With regard to the other PGMs it is important to appreciate that their availability is directly related to the output of platinum in the case of the South African reserves and of nickel in the case of the USSR and Canadian reserves. At current PGM prices it is not economic to mine and refine the ores for specific PGMs other than platinum.

GOLD

The major producers of gold are South Africa, the USSR, Canada, and the United States. The South African gold mining industry which currently produces 60% of the world output started in the 1870s and since then has contributed 40% of the total world production of 8.7×10^4 tonnes. There is little information available on gold production in the USSR and associated countries but sales from there account for about 25% of total gold supplies. In the United States about 60% of gold production comes from gold ores and the remainder as a co-product of copper and other base metal production. In the period 1973–77 world gold production has fallen from 1550 to 1450 tonnes, in part as a result of gold supplies being augmented from sales from Government stocks.

Not only is gold important to industry and the arts, but it retains a unique status among all commodities as a long-term store of value. It was long considered essentially a monetary metal, and most of the bullion produced each year went into the vaults of government treasuries or central banks. Since the late 1950s, however, the flow of gold to fabricators and private investors has exceeded monetary acquisitions and since 1968, when the major industrial nations agreed to abstain from further governmental acquisition of newly mined gold, the metal has become essentially a free-market commodity and prices have been free to adjust to supply and demand. Nevertheless, nearly half of all the gold that has been mined in the world is in government vaults, largely immobilized by agreements among the major industrialized countries. And while governments have attempted in recent years to reduce gold's traditional role in international monetary transactions, the policies and practices of many years' standing are slow to change.

The price of gold for the period 1968–78 is shown in Fig. 4. As far as the future supply/demand position is concerned the prevailing price has a bearing on the economic viability of marginal reserves and resources as well as on investor and industrial demand. For example, as a result of the substantial price increases which have occurred since 1968 some applications in the electronics industry have been superseded by cheaper precious metals such as palladium.

Industrial applications of gold
In addition to its unique appearance and rarity value, chemical inertness and tarnish resistance have, since ancient times, made gold unique as a treasured possession. In the second half of the twentieth century these chemical properties, coupled with other physical properties such as its high electrical and thermal conductivity, have made gold an essential item in many industrial products. The world consumption of gold by industry from 1973 to 1977 is shown in Fig. 5.

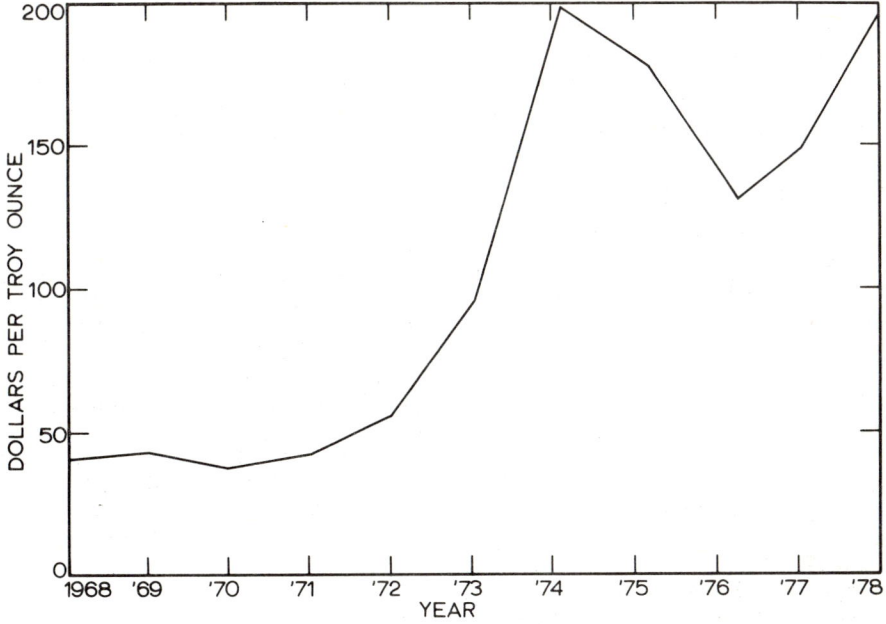

4   Average price of gold 1968–78

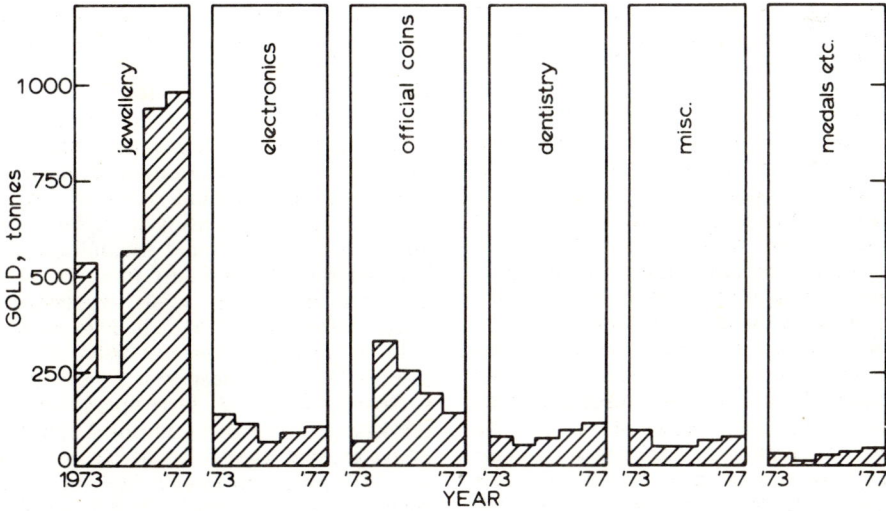

5   World consumption of gold 1973–77

Jewellery

The oldest use of gold, and its most important in terms of quantity consumed, is the jewellery industry. One reason for its extensive use is the belief that gold jewellery represents a store of value. However, its physico-chemical properties are also essential to its use in jewellery, since it is non-allergic, tarnish-free, and easy to fabricate. In 1977 the jewellery industry accounted for about 60% of the total gold consumed. Future demand for gold jewellery is expected to increase at approximately 2·5% per year.

Electronics

Of the industrial uses of gold the most important is in the electronics industry: for example, printed circuit boards, connectors, and keyboard contactors. Modern solid-state devices require components which will remain tarnish-free and stable throughout the life of the device as well as having acceptable electronic properties. Thus gold, together with other precious metals, is used in many applications. Initially quite large quantities of gold were used on individual components but more recently the industry has economized by developing selected-area plating and by using gold inlay and bimetallic strips for contact points. Miniaturization has also reduced the amount of gold used in this application. Gold alloyed with other precious metals as well as base metals is used in certain electrical components, for example slide wire potentiometers, where performance and long-term stability are essential.

Future demand for gold in this application thus depends upon trends in the electronics industry. On the one hand the use of electronic components containing gold is expected to increase, while on the other further economies may be made in the use of gold in these components. The electronics industry consumed 73 tonnes of gold in 1977, amounting to 4·5% of total consumption. While in a strategic sense this application for gold is important even a substantial increase would have only a minor impact on total demand. While some gold is recovered from electronic operations, the majority is distributed and lost.

Dentistry

A long-established application for gold, usually alloyed with other precious and base metals, is in dentistry. No significant changes in demand are anticipated in this area and as this constitutes a small percentage its effect on future availability is marginal.

Decorative uses

Gold is used to decorate a wide variety of pottery and glassware. It is applied in different forms depending on the substrate. An organic solution of gold is used for glass, gold metallizing pastes for china and pottery, and gold powder for transfers. Gold leaf is also used for decoration, e.g. in the lettering on the spines of books and interior and exterior decoration of buildings. The amount of gold used in each item is extremely small, therefore price and the impact on supply/demand are relatively unimportant.

Medals and official and fake coins

The collecting of medals and coins is a form of bullion investment. Consumption has increased since 1975 when the ban on private ownership of gold by US citizens was rescinded. It is the gold content of the medals and coins which is important, not the artistic quality. Reproduction coins and medals are especially popular in the Middle East. Commemorative medals are issued for anniversaries and special events, as for example the Olympic Games. It is interesting here to note, however, that for the next Olympic Games to be held in Moscow the Russians are reported to be minting a commemorative platinum coin.

In the next several years an increase is expected in the minting of official coins to absorb a strong demand for bullion coins for investment purposes. Any financial, economic political or military crisis aggravates uncertainty, which leads to a heavier demand for gold as a store of value. When investment interest is high, prices increase in order to bid sufficient supplies away from commercial markets or to attract additional supplies from above-ground stocks. In a strategic sense, as with the platinum metals, gold present in the form of medals and coins is unlikely to be lost and is ultimately convertible for important industrial uses.

Other uses

Other industrial uses for gold include bursting discs for plant protection and spinnerets for the manufacture of synthetic fibres. The turbine blades of jet engines are joined to the rotors using gold brazing alloys, and gold is used as a reflector of infra-red radiation in radiant heating and in heat-insulating windows for large buildings. In these applications gold makes an important contribution to energy conservation.

Secondary sources and recycling of gold

A considerable quantity of scrap is generated in gold manufacturing processes but because of the value of these waste materials nearly all of this is recovered. Refiners receive scrap in a variety of forms and the recovery process is dependent on batch size, average gold content, and impurities which may be present. Scrap dealers may carry out preliminary treatment and then transport it elsewhere for further treatment and refining. An interesting secondary source of gold and other precious metals in recent years has been identified in the sewage sludge originating from towns in America associated with the electronics industry. Secondary gold refining in the US has increased from 18 tonnes in 1973 to 32 tonnes in 1977.

Future supply of gold

Total world gold reserves and resources are estimated at 60 000 tonnes, half of which is reported to be in co-product form while half could be mined for gold alone if the price of gold makes this economically viable. In addition to mine resources the world's above-ground stocks of gold total at least 55 000 tonnes, of which two-thirds are in Government stocks.

World mine reserves and resources are therefore adequate to meet the projected demand till at least the end of this century. By then it is likely that world productions will be lower than at present and as much gold will be drawn from bullion stocks as is mined. It is expected that production of gold in South Africa will continue to decrease up to the year 2000 with new production coming from other parts of the world. These areas include

Siberia and Central and South America. It is possible that new gold fields may be found in areas such as the Witwatersrand Basin in South Africa and as yet unexplored areas of the USSR. An interesting recent development in South Africa has been the recovery of gold from the waste dumps of earlier mining operations.

SILVER

The major producers of silver are Mexico, the United States, Canada, Peru, Australia, and the USSR, which together account for 83% of total production at the present time. Currently 30% of silver is mined as a primary product, the rest as a co-product or by-product of other mineral production. Silver occurs in small amounts in most countries either as native silver or associated with other metals such as copper, gold, lead, or zinc. Sea water contains trace amounts of silver but it is not at present economic to extract. World

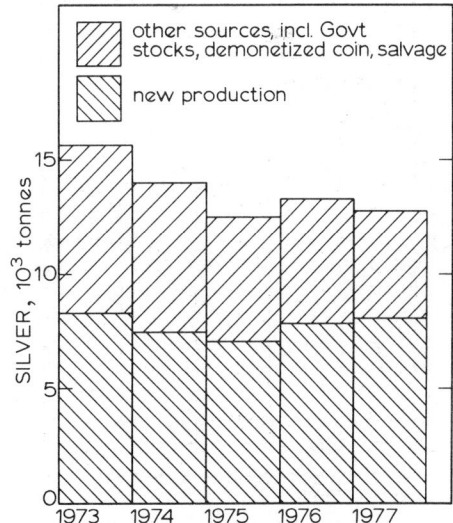

6 World silver supplies 1973–77 (excluding Eastern Bloc countries)

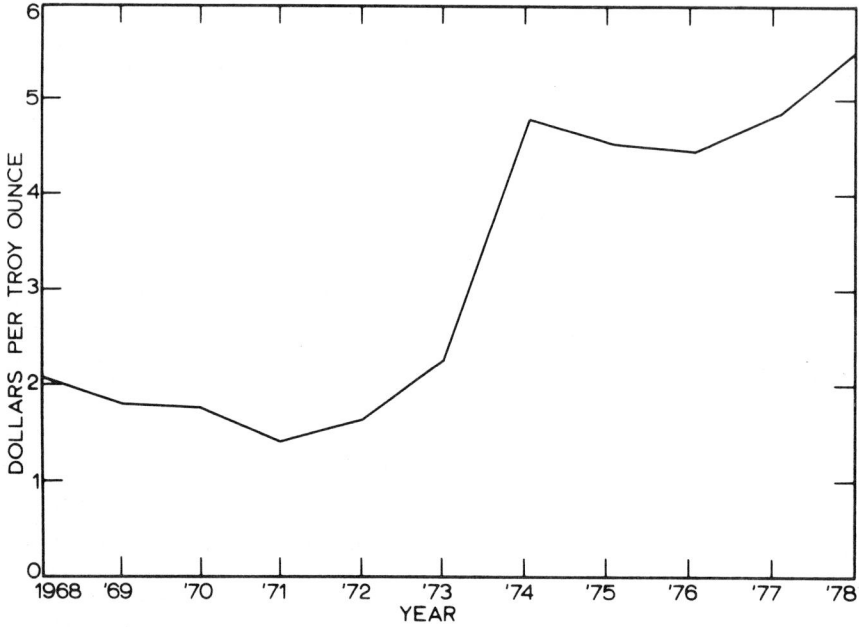

7   Average price of silver 1968-78

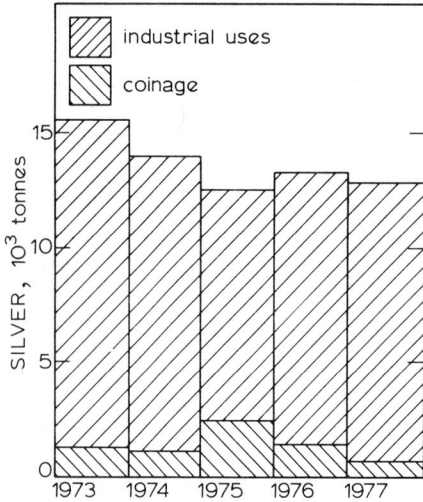

8   World silver consumption 1973-77
     (excluding Eastern Bloc countries)

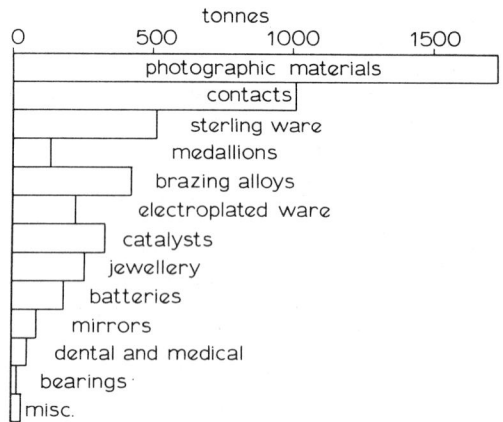

9   Silver consumption 1977 (USA)

demand for silver for industrial uses alone has greatly exceeded new production for the last ten years. Industrial consumers therefore look to above-ground accumulations as a source of supplies; accumulations which include government stocks, private stocks, demonetized coinage, and jewellery. Figure 6 gives the Western world supplies of silver for 1973–77, showing the relative amounts obtained from newly mined production and secondary sources. Above-ground accumulations in the United States alone total 10 000 tonnes in 1977 but India continues to be the major source of secondary silver. As with other precious metals price plays an important part in the supply/demand situation. The price of silver from 1968 to 1978 is shown in Fig. 7.

Current uses of silver

Refined shipments of silver are normally in the form of ingots which are converted to semi-manufactured products such as rolled and extruded bars, rods, wire, sheet, foil, powder, and pellets. These in turn go to manufacturers of silver products. One-third of silver goes to the manufacture of silver compounds. Figure 8 shows the 1973–77 world consumption of silver while the relative amounts consumed by each industry in 1977 in the US are shown in Fig. 9.

Photographic materials

Many silver salts undergo decomposition to give metallic silver when exposed to X-rays or to ultraviolet or visible light. The importance of silver halides in photography is not due to any unusual primary sensitivity to light but to the fact that they act as a catalyst for the development process. The paper on which the latent image is printed is also sensitized with a silver salt. The image definition obtained with silver salts has greater clarity than that obtained using other methods. Methods not involving silver are currently available and may become competitive in the future if the relative price of silver increases or silver were not available. These alternative techniques are presently only suitable for black and white photography and are not yet applicable to

colour reproduction. Photocopying technology which does not employ a silver-based system has only made a marginal impact on silver consumption because at present the technology is only suitable for low-definition work. It is possible that major changes in the use of other light-sensitive materials coupled with advances in computerized reproduction systems may in the future have a major impact in reducing the demand for silver in this application. However, over the last three years consumption of silver in photographic materials has increased to a record level. A substantial amount of silver used in this application is recovered and recycled.

Contacts

An ideal contact material for electrical switching should have the following characteristics: high electric and thermal conductivity, freedom from surface films, low cost, and resistance to electrical and mechanical wear. No single material has all these properties but silver and silver alloys are used for a large number of applications. Silver is prone to tarnish but in heavy-duty applications the voltage is high enough to penetrate the film and wiping action helps to break it down. Silver and silver alloys are the cheapest of the precious-metal contact materials and are thus widely used in electrical appliances including those in the home. In the future the number of domestic appliances is likely to increase but much of the silver used in contacts may be replaced by solid-state switching or composite contact devices which are expected to use less silver.

Brazing alloys

Two pieces of metal are brazed together by heating them up to or beyond the melting point of a filler metal which then effects a bond between them. Low-temperature brazing (600–850°C) is often called silver brazing because the filler metals are based on silver and copper. Silver brazing gives joints which can be as strong as the components joined and avoids excessive weakening of the metals brazed by the low temperatures used and the speed with which the joints can be made. Silver brazing is particularly adaptable to joining components of different thicknesses or heat capacities. It is widely used in pipe joints and making metal structures such as domestic appliances and motorcar components. Silver alloys are also used in soft soldering and in the electronics and jewellery industries.

Consumption of silver by brazing applications has remained fairly constant for the last ten years and no major changes are expected in the relatively near future. Yet changes in materials of construction aimed at energy conservation could lead to a reduction in the use of brazing techniques. Silver used for this application is largely dissipated and lost.

Other uses

A substantial amount of silver is consumed in tableware and jewellery. Silver is also used in batteries but the relative cost compared to other battery materials has limited this use to miniature batteries and to military and aerospace applications. Other uses include medicine, dentistry, mirror backings, bearings, and oxidation catalysts.

The use of silver in mirrors has expanded recently as a result of architectural trends and may expand further for solar energy applications. In all of these applications alternative materials are available in the unlikely event of silver not being obtainable. However, in most of these applications silver is recovered and recycled.

Secondary sources and recovery

Because demand for silver has exceeded primary production for several years, the recovery of silver from secondary sources is especially important. Indicative of this is that the quantity of silver recovered from secondary sources was some 25% greater than mine production in the US in 1977. The increased need for recovery has resulted in improved collection methods and recovery techniques.

Future supply of silver

World reserves of silver including above-grounds stocks as well as known resources are estimated to be capable of satisfying expected demands up to the year 2000. In the case of silver the extent to which government and private holdings, especially those in India, are made available for industrial use is partly determined by the prevailing price and also by government policy.

MERCURY

The major producers of mercury are the USSR, Spain, Italy, and China which together account for 65% of total world production. Mercury is sold in flasks, each flask containing 76 lb. Total production in 1977 was 240 000 flasks. Mercury is produced from its naturally occurring sulphide ore, cinnabar. Known reserves of mercury are small (3 200 000 flasks) and this led to speculation in the early 1960s that mercury would soon become scarce and all known reserves and resources would be worked out. This speculation did not forecast the concern about the toxicity of mercury which caused its uses to be re-examined nor the impact of new technology on its established major applications. These factors caused a reduction in demand and oversupply which was reflected in the price of mercury. Figure 10 shows the price of mercury (in dollars per flask) from 1968 to 1978.

Current uses of mercury

Because of the toxicity of mercury all its uses have come under increasing scrutiny to determine its possible effect on the environment. The best-known use for mercury is in thermometers, barometers, and pressure gauges but this accounts for only 10% of its consumption and in recent years solid-state instruments have been eroding this market. The major uses of mercury are in the chloralkali industry, in batteries, and in agriculture as pesticides and fungicides. However, other chlorine processes which do not use mercury have been developed and these are gaining

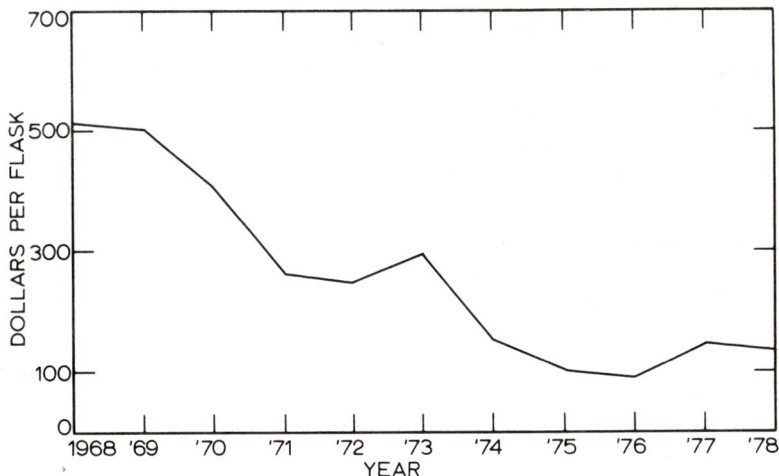

10   Average price of mercury 1968-78

favour for environmental reasons, other types of batteries provide adequate alternatives to mercury, and environmental considerations have also led to a preference for biodegradable pesticides.

Mercury has an important use in dental alloys, but gallium is now being considered as a possible alternative, again for toxicity reasons. Because mercury is a liquid at room temperature it will always find special uses, e.g. the 'lost wax' casting process, but where alternative materials can be used this is likely to occur for environmental reasons. In some of these applications, especially those least affected by environmental considerations, mercury can be recovered and collected.

Future supply of mercury
At current rates of consumption known reserves and resources of mercury would appear to be exhausted in 15 years. However, if the current trend of decreasing use in established applications continues (and the depressed state of the industry would confirm this) then this period could be extended. Despite the existence of recovery and recycle operations it is possible that by the year 2000 known reserves and resources could be exhausted.

BIBLIOGRAPHY

'Gold 1978' (Consolidated Gold Fields)

'Annual Bullion Review 1978' (Samuel Montagu & Co. Ltd)

'Gold 1976', *Mining Survey*, 1976, **82**, (3), 2

'Gold', *Mining Annual Review 1978*, p.27

'Industrial uses of gold', J. C. CHASTON: *Int. Met. Rev.*, March 1977, p.25

'The current gold view', (J. Aron Commodities Corp., October 1978)

'Minerals Yearbook 1976', Bureau of Mines (PGMs)

'Review of South African platinum shares', *Mining Research*, Ivor Jones Roy & Co., May 1978, 43–57

'Platinum': M.L. Doxford & Co., 1978

'Platinum metals', *Mining Annual Review 1978*

'Trends in mineral demand', D.G. MAXWELL: *Mining Survey*, 1975, 79, (4), 13

US Bureau of Mines Statistics

'The silver market 1978': Handy and Harman Annual Report

'Silver', *Mining Annual Review 1978*, p.31

'Mercury', *ibid.*, p.100

'Guide to non-ferrous metals and their markets', J. EDWARDS and P. ROBBINS: 1979, London, Kogan Page

Extracts from 'Encyclopaedia of chemical technology', Kirk-Othmer, 1966

'Platinum', MCP-22, US Bureau of Mines, September 1978

'Silver', MCP-24, US Bureau of Mines, September 1978

'Gold', MCP-25, US Bureau of Mines, October 1978

T. W. Farthing
and R. E. Goosey
Titanium

Titanium was the first of the new metals to be developed industrially and it remains much the most important. As a practical engineering material, its technology is effectively only about 25–30 years old. Considerable strides have been made in this time, not only in a better understanding of its basic metallurgy, but also in the improvement of process technology and production of all the usual wrought forms; alloy development; measurement of properties; and applications throughout many industries.

The occurrence of titanium is outlined and it is shown that titanium is an abundant metal, with estimated world resources amounting to nearly 800 million tons. The development of the titanium metal industry, stimulated initially by the invention of the gas turbine aero-engine, is described. Present markets and their possible growth are then considered along with the opportunities for and problems of changes in process and product engineering. It is predicted that applications for titanium will continue to expand and mature.

The authors are with IMI Titanium, Birmingham, UK.

OCCURRENCE

It is not often realised that although the titanium metal industry is still small, titanium is in fact the ninth most abundant element in the earth's crust and the fourth structural metal (after aluminium, iron, and magnesium). The principal minerals are ilmenite (50–60% TiO_2) and rutile (95% TiO_2). Some other titanium minerals are anatase (TiO_2), brookite (TiO_2) and perovskite ($CaTiO_3$). Estimated world reserves and resources are listed in Table 1; reserves being defined as those deposits considered workable with present technology and economics, while resources are those which are expected to be capable of economic recovery in the future. It will be seen that total resources amount to nearly 800 million tons.

Ilmenite is the commest mineral and the main source of titanium dioxide for the very large international pigment industry. This was established in 1910–20, using a sulphate process which was later replaced by a chloride technique in the late 1950s. This technology was developed in hand with the chloride extraction route for making titanium metal. The titanium industry is dominated by the demand for pigments. In 1976, for example, world production of titanium dioxide was equivalent to about 1·7 million short tons of contained metal, compared with only about 54 000 short tons of titanium as metal.

Table 1 Identified ilmenite and rutile resources world-wide (estimated, 10^6 short tons contained Ti)

	Reserves	Other resources	Total resources
ILMENITE			
North America	67	110	117
South America	1	5	6
Europe	47	22	69
Africa	10	150	160
Asia	39	100	139
Oceania	16	18	34
totals	180	405	585
RUTILE			
North America	1.5	4.9	6.4
South America	60.0	60.0	120.0
Europe	4.3	6.0	10.3
Africa	3.2	18.0	21.2
Asia	5.0	12.0	17.0
Oceania	6.1	1.6	7.7
totals	80.1	101.5	181.6

Almost all metal produced in the West comes from rutile. Australia is the most important source, from placer deposits (beach sands) in West Australia. Techniques are available for producing 'synthetic rutile' from ilmenite concentrates, from which high-quality titanium tetrachloride can be made for the production of metal.

DEVELOPMENT OF
TITANIUM METAL INDUSTRY

Titanium is fundamentally a highly reactive metal and is therefore difficult to extract. It forms a very thin film of titanium dioxide extremely rapidly at room temperature and at high temperatures will dissolve oxygen, with serious loss of ductility. A direct oxide reduction route is therefore not practical, and the foundations of a practical extraction route had to wait until the development of halide metallurgical processing. The most important contribution came from Dr W. J. Kroll in the late 1930s, when he successfully reduced titanium tetrachloride with magnesium to produce titanium 'sponge'. It is

interesting to note that $TiCl_4$ was reduced by sodium nearly 30 years earlier. Sodium and magnesium reduction of $TiCl_4$ are still the basic routes used today.

The stimulus to the development of titanium metal came originally from the aircraft industry. The aircraft gas turbine had only just been invented and, because of its need for high strength-to-weight, creep-resistant materials, offered an attractive market for titanium. The technology was developed rapidly during the late 1940s and '50s to meet this need, and has continued up to the present day.

As already mentioned, the original driving force came from the rapid post-war growth in air transport. Demand was mainly for alloys and required the development of new and often complex materials to be used at the limit of their properties. Even as recently as the early 1960s the aircraft market dominated the picture and represented around 95% of the engineering demand for titanium. Because of the cyclical nature of the aircraft business, this imbalance was not commercially attractive and it underscored the need to search for new markets. Broadly speaking, the work was based on the outstanding corrosion resistance of pure titanium and it started very early in the life of the metal. With some notable exceptions, it has been mainly a case of continuous application of research and development. Considerable effort was involved in developing the real value of the metal and in providing new designs and data, fabrication techniques and training for plant builders to overcome their reservations about this 'new' space-age metal. The work has succeeded to the extent that the titanium market in Western Europe is now about 50% aerospace and 50% chemical and general engineering.

This development of new products and markets had the enormous advantage of working from a firmly established process base formed by the initial work for aerospace. The extraction of the raw metal had been established on a production scale; new vacuum arc melting and fabrication techniques had been worked out; the basic metallurgy and properties were better understood.

It is worth emphasizing that progress would have been much slower without corresponding developments in other sectors of industry. This applies not only on the user (i.e. the product) side, but also to plant and process equipment. This complex interaction of different industries, increasingly on an international scale rather than simply in the UK, is central to the theme of this meeting. Our lack of indigenous metallurgical resources obliges us to look to our place internationally. The problems that this presents should not be minimized, in the light of our inability to agree a UK domestic energy policy.

MARKETS FOR TITANIUM

Aerospace

Some of the main applications for titanium are listed in Tables 2 and 3. The most important uses in engines are for compressor blades

and discs, and the first commercial application of titanium alloys in the UK was as blades in the Rolls-Royce Avon, Conway, and Tyne engines. The many uses which followed are given in the tables. The development of creep-resistant alloys is illustrated in Table 4. Hylite 50 (originally invented by Jessop Saville, later adopted and developed as IMI 550) was used for discs in the Rolls-Royce Pegasus engine. Along with IMI 318 blades, this achieved valuable weight saving in the Hawker Siddeley P1127 vertical take-off aircraft (later the Harrier). Titanium was selected in 1961–62 for the Bristol Siddeley Olympus 593 engine for Concorde. This uses about 4 tonnes of mill products per engine: IMI 550 for the discs and rings; IMI 318 for blades; and IMI 230 for guide vanes, air casings, etc.

A welded titanium compressor drum was introduced in the Rolls-Royce Adour engine,

Table 2   Titanium aerospace developments

Year titanium introduced	Engine	Major titanium applications and alloys used*	Associated airframes
1954	Avon	315 compressor blades	Comet, Caravelle, Hunter, Canberra, Lightning
1959	Conway	315 blades, 230 by-pass duct	707, DC8, VC10
1960	Tyne	315 blades	Vanguard, Belfast, Atlantic, Transell C160
1960	Pegasus	Hylite 50 (IMI 550) discs and blades, 318 blades	P 1127, Harrier
1961	Olympus 320	Hylite 50 discs and blades,	TSR2 (cancelled 1964)
1962	Olympus 593	Hylite 50 discs and blades, 318 blades, 230 intermediate air casing	Concorde
1962	BS100	Hylite 50 discs, 318 blades	HS 1154 (cancelled 1964)
1962	Spey -1, -2, -25 series Spey Mk 101	679 blades, 230 by-pass duct	Trident, BAC 1-11 Buccaneer
1964	Spey Mk 201	679 discs/blades, 230 by-pass duct	Phantom
1964	RB172	684 welded HP compressor drum (later 685),	Jaguar, Hawk, Mitsubishi T2
	Adour	155 by-pass duct	
1965	BS360	684 discs, later replaced by 318 for axial discs and 550 impeller	Lynx
1968	RB211	6A1-4V rings, fan, LP discs, 685 HP discs, 230 casing, 318/685 blades	Tristar, Boeing 747
1968	RB199	Used extensively for both rotating and static components (230, 318, 685)	Tornado

*The numbers refer to IMI alloy nomenclature

Table 3 European aircraft projects using IMI titanium

Aircraft	Manufacturer	Engines	Airframe applications	Engine applications
Tornado (MRCA)	BAC, UK/MBB, West Germany/ Aeritalia, Italy	RR/MTU/Fiat RB199, 2 off	IMI 230 sheet fabricated keel IMI 550 flap and slat tracks IMI 230 and 318 for skinning	IMI 230, 318 and 685 used for both rotating and static components IMI 318/Tital castings being evaluated
Jaguar	BAC, UK/Dassault-Breguet, France	RR/Turbomeca Adour, 2 off	IMI 230 sheet fabricated keel IMI 550 slat and flap tracks, brackets	IMI 685 welded HP drum IMI 155 by-pass duct IMI 318 LP discs and blades
Hawk	HSA, UK	Adour, 1 off	Small amount CP titanium	See Jaguar above
Harrier	HSA, UK	RR Pegasus, 1 off	IMI 551 and 318 components and fuselage skinning	IMI 550 and 318 discs and blades
Lynx helicopter	Westland, UK/Aerospatiale, France	RR, GEM	IMI 318 tail rotor hub and gear box fittings	IMI 318 and 550
Puma helicopter	Westland, UK/Aerospatiale, France	Turbomeca TURMA, 2 off	IMI 318	No titanium
Alpha Jet	Dornier, West Germany/ Dassault-Breguet, France	Franco-German Larzac, 2 off	No IMI titanium	IMI 685 welded HP drum IMI 318 LP discs and blades
Concorde	BAC, UK/Aerospatiale, France	RR Olympus 593, 4 off	IMI 230, 318 and 680 used in engine nacelles	IMI 230 casings IMI 550 discs IMI 318 blades
A300B	Aerospatiale, France/Deutsche Airbus, West Germany/VFW-Fokker, West Germany/HSA, UK/ CASA, Spain	SNECMA/GE CF6-50, 2 off	IMI 500, 318 and 230 in wings, engine pylons	No IMI titanium
Tristar	Lockheed, USA	RR RB211-22 or -524, 3 off	No IMI titanium	IMI 230, 318, 685 and 550 used for rotating components, fan casing and mounting brackets

Table 4 Development of IMI creep-resistant alloys

Alloy	Year of introduction	Maximum operating temperature, °C
IMI 550	1958	400
IMI 679	1959	450
IMI 684	1964	475
IMI 685	1968	520
IMI 829	1976	550

made from IMI 685 alloy. This combined high creep strength in the range 450–520°C with high fracture toughness and good weldability. This same alloy is used in the compressors of the Rolls-Royce RB211 and RB199 engines.

The most recently introduced creep-resistant alloy is IMI 829, developed specifically for blades and discs operating at temperatures up to 550°C or even as high as 600°C in low-stress applications. Like the 685 alloy, IMI 829 is fully weldable.

The use of titanium in airframes in the UK has not been as extensive as in engines. Early applications consisted of unstressed parts such as fire-walls, bulkheads, and so on in commercially pure metal. IMI 230 (2½%Cu) is used today in the Jaguar, Tornado, and Concorde. About 16 tonnes of mill products are used on each Concorde for such things as nacelle components, ducts, and brake assemblies. IMI 550 has become an established structural alloy for such parts as flap and slat tracks and engine brackets in many other aircraft including the Airbus, Jaguar, etc.

Non-aerospace

Most non-aerospace outlets use commercially pure titanium on account of either its outstanding corrosion resistance or its unique electrochemical characteristics. A selection of uses is indicated in Table 5. It finds wide use, for example, in metal finishing for anodizing jigs, as well as for general construction. An important use has been developed in the metal refining industry, especially as starter blanks for electrolytic copper. Another major

innovation has been in the chlorine industry. Titanium is passive under all the corrosive conditions found in the production of chlorine, and it is therefore used extensively for heat exchangers, valves, and precipitators. Its most important use in the chlorine industry, however, is in dimensionally stable anodes (DSA). These consist of titanium coated with a very thin layer of precious metal; the titanium is passive in the brine but passes current via the PM coating. The titanium is a more satisfactory engineering material than the traditional graphite (it is cleaner and sustains less wear in service, for example). Even more important, however, is the fact that, because of the lower chlorine overpotential of the PM coating and a reduction in the solution resistance from the smaller anode–cathode gap, PM-coated titanium is fundamentally a more efficient electrode. This leads to major savings in power consumption, and the chlor-alkali industry is now a major market for titanium.

Important applications have also been developed in processes such as those involving acetic acid, benzoic acid, malic acid, amines, urea, terephthalic acid, vinyl acetate, and ethylene dichloride. The justification here for using titanium can stem either from the corrosive attack of the reaction products themselves, or from aggressive catalysts.

The oil and petrochemical industries have also developed into significant markets for titanium for heat exchangers, process coolers, etc. Offshore oil rigs especially provide conditions which highlight the remarkable corrosion resistance of the metal. Sea-water cooling is used, with sulphide-contaminated products. Titanium completely resists both of these and provides a 'fit and forget' service.

Another important outlet for titanium is in steam condenser tubes, based on its outstanding corrosion resistance to polluted, brackish or sea waters. Much thinner-walled titanium tubes can be used than tubes of traditional copper alloys; only 0·7 mm, for example, compared with 1·2 mm for copper alloy. This greatly improves the economics of using titanium on a first-coat basis. On a whole-life basis, titanium avoids the high cost of unscheduled outages on account of premature tube failure.

Table 5 Applications of IMI titanium in chemical and general engineering

Industry	Environment	Alloy used	Types of plant	Reason for using titanium
1 Chemical	Organic acids; chlorides/bromides; chlorine compounds; chlorinated solvents; nitric acid; chromic acid; fertilizers (urea, ammonium nitrate etc); sulphuric, hydrochloric and phosphoric acids	CP Ti, IMI 260, IMI 261	Loose lined, solid and clad plate vessels; tubular and plate type heat exchangers; mixers; pumps; valves and pipework; dimensionally stable anodes for chlorine production	Excellent corrosion resistance in wide variety of oxidizing and neutral solutions, particularly in chloride environments. Good corrosion resistance in non-oxidizing solutions in presence of inhibitors, with anodic protection or as IMI 260 or IMI 261. Unique electrochemical properties for anode use
2 Paper pulp and textile bleaching	Chlorine dioxide; sodium hypochlorite	CP Ti	Tanks; J-boxes; heat exchangers; heating elements; rollers	Good corrosion resistance in chlorine-containing compounds such as chlorine dioxide and sodium hypochlorite
3 Metal finishing	Nickel, chromium, copper plating solutions; sulphuric acid anodizing solution	CP Ti	Tanks; heat exchangers; heating coils; immersion heaters; anode baskets; anodizing jigs; anodes	Good corrosion resistance in most metal finishing solutions. Unique electrochemical properties as anode basket, jig or anode material
4 Metal refining	Various, mainly sulphuric acid-based solutions	CP Ti	Heat exchangers; coils; starter blanks; cathodes; anodes	Good corrosion resistance; non-stick properties; electrochemical properties
5 Power generation	Seawater, brackish and polluted estuarine waters	CP Ti	Steam condensers; ancillary heat exchangers	Good corrosion resistance to seawater and other polluted, naturally occurring waters
6 Desalination	Seawater and various sodium chloride brines	CP Ti	Heat exchangers for multi-stage flash distillation units	Good corrosion resistance to seawater and sodium chloride brines
7 Steel	Hydrochloric acid and ferric chloride; hydrochloric acid gas	CP Ti, IMI 260	Various plant items	Good corrosion resistance
8 Nuclear	Nitric acid	CP Ti	Vessels for nuclear fuel reprocessing	Guaranteed good corrosion resistance in nitric acid
9 Surgery	Body fluids	CP Ti, IMI 318	Prostheses; joint replacements; bone plates; screws, pins, etc.	Good corrosion resistance and compatibility with body fluids
10 Marine	Seawater	CP Ti, IMI 318	Tubular and plate type heat exchangers; scrubbers; sonar; yacht fittings, etc.	Good corrosion resistance; high specific strength
11 Oil production and refining	Seawater,; brackish water; sulphide containing hydrocarbons	CP Ti	Tubular and plate type heat exchangers	Good corrosion resistance to seawater and to hydrocarbons
12 Engineering	Various	CP Ti, Ti alloys	Various	Good specific strength

FUTURE
PRODUCT DEVELOPMENT

Titanium is now an established engineering material. Its balance of properties and characteristics is unique and guarantees its place in the portfolio of materials open to the design engineer. No other non-ferrous material finds its way into such a range of applications. The size of the industry is still small, but there will be continued growth in those applications which have already been opened up.

The most obvious example is aerospace, which is facing a massive programme of re-equipment. In Europe the principal civil airliner programme today is the European Airbus, for which British Aerospace is supplying the wings. British Aerospace itself is launching the new small feeder aircraft (BAe 146) which uses titanium in the airframe.

The most important aero-engine for the UK, and one which incorporates a significant weight of titanium, is the Rolls-Royce RB211 family. The original major use for this was in the Lockheed TriStar, and the long-range TriStar 500 has just been ordered by Pan-American. The RB211 engine is also being specified by more airline operators for the Boeing 747, and offered by Boeing as the first-choice launch engine for their new twin-engine 757 aircraft, which will take the place of the 737.

On the defence side, titanium is widely used in both the airframe and Rolls-Royce RB199 engines of the Panavia Tornado being produced for the British, West German, and Italian air forces. The Rolls-Royce Adour engine, which utilizes titanium in the compressor, has been very successful in the Jaguar and Hawk aircraft. This engine is being built under licence in Japan for a Japanese-designed aircraft, and will also be built in India for their Jaguar requirements.

All these programmes will require considerable and increasing volumes of titanium well into the 1980s. A similar position exists elsewhere in the world's aircraft industry. This is not surprising, for it is difficult to see any slackening of the demand for air travel; it is vital and growing, not only for business purposes but also for leisure. The advent of cheaper air travel is just one element in the growth forecasts on which the airline re-equipment programme has been based.

Furthermore, it seems likely that if the element of cost involved in leisure air travel continues to fall, and if the other holiday costs continue to rise (as they are doing in Western Europe) then there will be an increasing tendency to take holidays further afield. An argument that is sometimes advanced in opposition to this is the increasing cost of fuel which, it is claimed, will reduce the growth in air travel by forcing up fares. However, the use of titanium, both in its simple weight-saving role and also in improved engine performance, is directed at this very problem; as fuel becomes more expensive, and aircraft become more expensive to fly, the case for titanium becomes even stronger.

Alloy metallurgy is still dominated by aerospace needs but, with few exceptions, there has been comparatively little development of new alloys in the last ten years, certainly by comparison with the initial phase in the life of titanium. Considerable effort has of course continued to be devoted to alloy metallurgy, but the main advances have been more in the interplay between thermo-mechanical variables and the structure and properties of existing alloys than in the invention of completely new compositions. In consequence, there is now a better understanding of the effects of heat treatment on microstructure and on such properties as fracture toughness.

The search for even-higher-duty materials to improve the efficiency of aero-engines might well bring the development of new alloys and compositions more into the forefront than it has been in recent years, and there are signs of this in the work on titanium aluminides. One feature of the alloy scene is that — as understanding of the basic metallurgy improves and becomes more detailed; as demands increase, not just for increased properties but also for improved quality in performance; and as close co-operation continues between aerospace design engineers and metal producers — the task of inventing and developing new alloys and bringing them into successful production becomes more difficult. Aerospace is one of the most demanding areas of technology and, perhaps with rather a different balance than in the past, we need not only to continue with research but also to consolidate and build on past gains.

The most notable work in recent years has been, and for the foreseeable future will be, in product development outside the aerospace area and in process development. A number of striking examples of new products in the chemical and general engineering markets have already been touched on. The invention of dimensionally stable anodes for the chlor-alkali industry, for example, was a major development and technical advance. The increasing application of tubing called for precision thin-walled material, in both the seamless and the welded forms.

Wider growth, and breakthrough into new markets, could depend to a large extent on the users' attitude to price. Titanium is sometimes considered to be an *expensive* metal, and for applications for which it is neither suitable nor necessary this is no doubt true. However, even in first-cost terms, the cost of titanium must be considered on the basis of volume rather than mass. It scores on this account because of its low density; titanium plate, say, 4in thick is cheaper than Monel 400 and very much cheaper than Hastelloy C-276. The advantage is often much more dramatic when the 'cost of ownership' is taken into account, i.e., not just the initial cost but the savings which can be made by using titanium. In chlorine production, for instance, the use of DSA anodes provides huge savings in energy consumption, owing to the lower chlorine over-voltage and the smaller anode–cathode IR drop. In steam condenser tubes the use of titanium obviates the cost of retubing during the condenser life and (perhaps even more significant on account of its unexpectedness) the cost of unscheduled stoppages from leaking condenser tubes. Modern power plant boilers are becoming even less tolerant of minor contaminations in the water, and it is vital to ensure this integrity of the condenser at all times. To date, all the titanium which has been used commercially in steam condensers has been plain tubing; the use of extended surface tubing for improved heat transfer could strengthen its position still further.

The possible application of titanium in steam turbines, for example in blades, has been studied for many years. In the authors' own organization, for example, titanium alloy blades have been successfully used for almost ten years in a steam turbine. This has resulted in an increase of about 15% in the output of the machine. Similar trials have been conducted in many other countries, where titanium blades have been satisfactory for up to 20 years in the low-pressure stages of steam turbines as an alternative to 12% chromium steel. Commercial interest can be expected to grow in this application. Another feature of the use of titanium in the chemical industry is that it has facilitated the development of processes which would prove extremely difficult without it. An example is the production of terephthalic acid, which uses halide catalysts.

Expanding use of titanium can be expected in such areas as desalination, for which its outstanding corrosion resistance is eminently suitable in the MSF process. The recovery of 'waste' water by recycling will become more important as water resources become scarcer. One way in which this could show itself — as indeed it already has in the UK — is the increasing use of sea or estuarine water for cooling, with the siting of more and more industrial plant on the sea coast, which will demand plant constructional materials which can resist sea-water corrosion. Indeed, the sea itself is a major natural resource which, one might argue, will be increasingly exploited. This is seen in such projects as the ocean thermal energy conversion (OTEC) effort in the USA, and even in sea-bed mining for manganese nodules. The offshore oil industry, another sector in which titanium is already widely used, can be expected to become more widely developed than now, in the search for new oil resources.

Apart from specialized and relatively small volume uses, such as in racing cars, titanium has so far made comparatively little penetration into the general mechanical engineering market. This position might change, the basic reason being the same energy-saving factor which is behind the use of titanium in aerospace. This is not to predict, for example, that car bodies will be made in titanium, but it is not unreasonable to expect a widening range of specialized uses such as springs, connecting rods, suspension arms, and the like. Another factor, which is already evident in other applications for the metal, is the engineer's growing confidence in using titanium, undoubtedly stimulated by the wealth of successful experience in its well-established, wide-ranging areas of application. This is of course a characteristic of the technological 'learning curve' found with all new developments.

FUTURE
PROCESS DEVELOPMENTS

Much of the recent work on improved processes has been in the aerospace field and aimed broadly at improving the level of material utilization, i.e. the 'fly-weight'. Yields of airframe components especially can be very low when large, oversize forgings have to be machined down to a small size and weight; quite apart from the costs of the actual machining itself. Another major cost area lies in the actual assembly of airframe components, both in the cost of the fasteners and the labour cost and time of assembly. These have led to a growing volume of work on such things as:

(a) *Isothermal forging*

This is a technique of forging wherein the forging press tools are heated as well as the forging metal stock itself. The thermal chill during normal forging with 'cold' tools is a major reason why fine detail or close-to-form forgings cannot be produced. Preheating of the press tools minimizes this chill, and facilitates the production of close-to-form parts.

It is expensive of course, since the press tool materials must themselves have high strength at the working temperature. Nickel-base alloys, and even possibly molybdenum-base materials of the TZM type, are usually required; these are expensive in themselves, and are not always easy to make into tool shapes. The extra cost of this item must therefore be set against the saving from better utilization of the metal being forged.

A derivative of this technique which has been used consists of finishing a conventional forging isothermally. Savings of around 20% in the finished part have been claimed for the technique. One feature that is still not easy to provide is a final 'net shape' which does not require final machining, the latter being usually a major cost compared with rough machining.

(b) *Superplastic forming*

Under the proper conditions of microstructure, temperature and pressure, titanium alloys become superplastic, exhibiting tensile elongations of 1000% or more without necking. This means that the metal will flow much more readily into the die cavity and more precisely reproduce fine detail, including sharp corners and radii. This has obvious advantages in material utilization over, say, machining. It can also be combined with another technique which is readily adaptable to titanium, namely diffusion bonding. This relies on the fact that clean titanium will readily pressure-weld under the right conditions of temperature and pressure.

The combination of superplastic forming and diffusion bonding can therefore reduce cost not only because of better material utilization, but also by reducing the cost of assembly. The actual extent of saving depends of course on the complexity of the part being made, but significant cost and weight savings have been claimed.

(c) *Powder metallurgy*

This classic approach to improving material yields is being studied for titanium. Many 'standard' techniques have been mooted, including cold compacting and sintering, vacuum hot pressing and, for complex shapes, hot isostatic pressing. The latter enables a complicated shape and different section thicknesses to be densified by equal amounts. This is certainly true in theory, but it depends crucially on achieving and maintaining the same powder characteristics and size distribution across the initial pre-compacted form. Without this, either the final density or the section thickness will vary.

Further problems arise in respect of the powder itself. There are a number of different methods of producing powder, such as attribution, hydride–dehydride, and melting; the latter consists of melting off the tip of a rapidly rotating electrode (the REP process). Each of these produces different characteristics. Hydride powder, for example, is angular, while REP material is spherical. Furthermore, since titanium is a highly reactive material, it will readily pick up surface contaminants if not handled extremely carefully.

These are just some of the techniques that are being studied. Before becoming widely applied, certainly in critical parts, a great deal of property evaluation of the finished products will be needed. This will ensure that quality and reproducibility and yield in the process itself are satisfactory; and that there are no unforeseen problems or deficiencies in the final product.

ENERGY AND RECYCLING

Two specific themes of this meeting are the influence of energy on fabrication processes, and the question of designing for recycling. As far as energy is concerned, thermodynamics dictate that in fundamental terms titanium will always be more energy-hungry than many other metals. However, it might be more to the point to say 'energy-rich' since, compared with many other engineering materials, titanium provides a much lower cost-of-ownership. It is indeed expensive to extract, but in return it gives a much better performance. This is the case with dimensionally stable anodes, titanium starter blanks for metal refining, heat exchange tubes for many applications, chemical plant and so on; and it is, of course, true for the largest single market, namely aerospace. The modern gas turbine develops as much as 50 times the thrust of the original Whittle engine, and compressor exit temperatures have doubled from around 250°C to over 500°C. These engineering advances would not have been possible without matching improvements in materials, including titanium. Titanium provides a pay-off simply in facilitating the development of more efficient engines and aircraft.

In general process terms, the search for improvement in titanium processing will follow the lines applicable to all materials, which will be accentuated by the increasing cost of energy. It has, for example, always been a continuous problem for the metallurgist to improve first-time yield and, as already indicated, new approaches are being followed in this work. Arising from its metallurgical properties, a general guideline in increasing the yield in titanium processing is to carry out

the operations as quickly as possible and at as low a temperature as possible. These points have already been reflected in equipment currently used for primary forging.

It is possible that technical constraints imposed by the market may limit the freedom to change processes. For example, aerospace requires materials of the highest and proven quality, and some of these requirements would be difficult to change without overwhelming evidence that the quality had not been reduced by the change.

The question of designing for recycling (i.e. making the original design more suitable for recovery at the end of useful life) is often raised at contemporary discussions about resources. It is a valid point if one could argue that the 'cost of ownership' also includes the value of disposal, but the question is: can we? Is not the dominant and primary requirement facing most engineering designers and suppliers that of *use* — not the question of scrapping it when the artefact ceases to be of use? Certainly, if in designing for recycling, one could cheapen the first cost, then the product should be redesigned simply to achieve this. If, however, in designing for recycling the first cost is increased at all significantly (and perhaps even the operating cost-of-ownership), then where is the guarantee that this cost will be recovered at the end of the lifespan, in cash terms realistic enough to be included in the original DCF appraisal? The objective may be arguable in terms of ecology and conservation of resources but the practical, commercial task is the optimum deployment of resources (including finance). It is a difficult enough task, for example, to argue a net benefit in the operating cost-of-ownership when this significantly increases the first cost, without the additional complication of the value of the obsolete plant at the scrapping stage.

CONCLUSIONS

The original stimulus for the development of titanium came from aerospace, and this will undoubtedly continue as the largest single market. Its importance is likely to grow with the designer's increased demands on the metal as aircraft and engines continue to improve. These advances will call for corresponding effort from the metal-producing industry. Titanium has an important advantage over many other engineering materials in being abundant and found in many parts of the world. Scarcity or political uncertainty should not therefore be a problem.

The technological base will be strengthened by work aimed at making more effective use of the metal, and this should also help in furthering the progress which has been made in penetrating other markets outside aerospace. As the service conditions and requirements become more severe—whether, in the search for improved efficiency, these be in the reaction conditions of new chemical plant and processes or the growing financial cost of unscheduled down-time—there seems little doubt that applications for titanium will continue to expand and mature.

SESSION C
Influence of design factors on future utilization of metals

R. L. P. Berry
and L. Whalley
Recycling of metals

There is no doubt that in the United Kingdom recycling of metals is very important to the economy and that the secondary metals industry performs a vital service. In a recent survey by the British Reclamation Industries Confederation it was revealed that the largest of the participating bodies, the British Scrap Federation, recovered 10·3 million tonnes of ferrous scrap worth £345 million and 322 thousand tonnes of non-ferrous metals valued at £109 million during the 12 months to June 1978. An earlier survey by the British Secondary Metals Association estimated that its members handled non-ferrous metals worth

The importance of the recycling of metals to the UK economy is reviewed. The structure of the ferrous and non-ferrous secondary metal industries is described, and the factors which determine whether or not a particular metal-bearing waste can be recycled economically are considered. Estimates are given of the quantities of low-grade metal-bearing wastes which are currently lost to the economy, and the prospects for improving the recycling efficiency of metals are discussed.

Dr Berry is Director of the National Anti-Waste Programme; Dr Whalley is at the Warren Spring Laboratory of the Department of Industry in Stevenage, Hertfordshire.

£850 million. It is reasonable to assume that a proportion of the sales of non-ferrous metals by BSF members was to BSMA members, but the value of scrap metals handled by the UK reclamation industry must be more than £1100 million/year. This figure greatly underestimates the total value of scrap to the economy because of the large quantities which are generated in-house or sold directly to the mills.

Consumption of ferrous scrap in 1973 and 1977 is shown in Table 1. It should be noted that the total scrap recovered from UK sources in 1977 was considerably in excess of the quantity sold by the members of the British Scrap Federation. This table also illustrates another important point: scrap and waste arisings relate to industrial activity and the amounts available and the annual demand vary widely. The fall in demand of almost 3 million tonnes in 1977 compared with 1973 reflects the current problems being experienced by the steel industry.

The importance of metal recycling is usually illustrated by statistics for amounts recovered as a percentage of total consumption. This is shown in Table 2 for ferrous metals and the five most commonly encountered non-ferrous metals. Since the available figures for scrap recovery do not differentiate between new (process) scrap and old (post-consumer) scrap, there is a problem in identifying areas where

Table 1   Recovery of ferrous scrap

	Quantity recovered, 10^3 tonne	
Use	1973	1977
For castings:		
steel scrap	1 309	1 242
other	2 732	2 390
Steel- and ironmaking:		
blast furnaces and sinter plant	785	636
steelworks and steel foundries	13 862	11 222
refined ironworks	205	195
Net exports (exports minus imports)	517	838
Scrap recovered from UK sources	19 410	16 523

recovery could be improved. The amounts of residues and new scrap available depend on current industrial activity, while the amount of old scrap available is a function of past industrial activity and obsolescence rates. The long-term use and increasing consumption of many of the metals mean that the amount recovered as a proportion of current consumption is unlikely to rise above 60% and is considerably lower for many metals, particularly those with substantial dispersive uses. Of course, if major discontinuities in supply bring about substitution this situation could change.

THE SECONDARY METALS INDUSTRY

The scrap industry consists of a number of tiers, independent but working closely together. Small collectors gather waste from local industry and domestic sources; the metal constituents are passed to larger metal merchants who combine them with directly purchased works clearances. During the last few years many of the smaller firms have disappeared and the trend has been towards the formation of larger fully integrated units with considerable resources to invest in modern plant. Another tendency has been for

Table 2   Metals consumption in the UK

	Total consumption, 10^3 tonne		Amount from scrap, 10^3 tonne		Amount from scrap, * % of consumption	
	1973	1977	1973	1977	1973	1977
Copper	718	643	271	207	38	32
Zinc	395	315	90	70	23	22
Lead	364	290	227	174	62	60
Tin	18·4	14·9	2·7	3·6	15	24
Aluminium	687	589	200	171	29	29
Ferrous						
(iron and crude steel)	26 594	20 411	14 853	12 052	56	59
(from castings)	3 445	2 795	4 098	3 632		
Ferrous all products	30 039	23 206	18 951	15 684	63	67

*Excluding net exports of raw and ingoted scrap

new scrap to be returned directly to the mills; this has become possible as a result of greater awareness on the part of users of the value of scrap provided it is clean and segregated.

FERROUS METAL SCRAP

Steel production and the balance between the various steelmaking processes determine the demand for ferrous scrap. The proportions of scrap used vary from that of electric arc furnaces with their 100% scrap charge to the 35–60% used in the obsolescent open-hearth furnaces and the 30% used in basic oxygen steelmaking. In practice the ferrous scrap available is rarely the optimum size, shape or density for steelmaking. Modern ferrous scrap recovery depends very largely on the use of sophisticated handling equipment, either to compact light scrap into bulks of manageable size and density or to reduce the size of heavy scrap by shearing or fragmentizing. The latter process, which has assumed increasing importance particularly in supplying feed for mini-mills, is based upon the use of a hammer mill to reduce the scrap to fist-sized lumps. The most sophisticated machines use a stream of air (advantage is often taken of the strong air currents generated in the hammer mill) to remove dirt, plastics and rubber. The non-ferrous content of the scrap can be reduced by installing a magnetic separator after the fragmentizer.

The chemical characteristics of ferrous scrap are even more important than its physical form. Most serious in terms of steel quality is the presence of elements which cannot be eliminated during steelmaking and whose unintentional presence may impair the properties of the finished product. These 'residual' elements include copper, tin, nickel, chromium, molybdenum, sulphur and phosphorus. Maximum permissible concentrations of these residual elements depend upon the grade of product. In strip mill products, where the effects of impurities on surface quality, strength and formability are vitally important, maxima of 0·03%Sn, 0·1%Ni and 0·1%Cu are not uncommon. At the other extreme, a typical specification for reinforcing bar might be 0·1%Sn, 0·3%Ni and 0·3%Cu.

These limitations on residual elements have an important bearing on the acceptability of the lower grades of scrap. For example, the only UK market for ferrous metal extracted from household waste is in refined iron-making, where the tin content is beneficial for the manufacture of certain grades of cast iron. There is concern in the steel industry about the steady rise in the levels of residuals in steel, which can only have come from merchant scrap. The problem has been contained by increased dilution with iron in integrated steel plants, and careful control of scrap quality is essential. To remain viable the scrap industry has transformed itself from a labour-intensive to a capital-intensive industry, but a by-product of this change has been a decline in collection and dismantling efficiency and increases in residuals in steel.

NON-FERROUS METAL SCRAP

Non-ferrous metal scrap can be sold for secondary ingot production, castings and other direct uses or to smelters and refiners for production of secondary metals. Direct users require well-sorted, high-grade scrap and suitable material usually changes hands at premium prices. Lower-grade material and mixed alloys can be sold to secondary refiners and in this case the price received depends upon the recoverable metal content. Because of relatively low metal prices since 1975 it has often been more attractive for scrap processors to upgrade some of the better qualities of refinery grade scrap to direct-use specification rather than to sell for refining. Complex alloy scrap presents special problems and is often sold to specialist merchants.

Alloys purchased must be identified reliably and rapidly. Material positively identified and carefully segregated as one particular alloy can be sold for remelting to produce prime ingots of the same alloy. If it cannot meet the demanding specifications for this end-use it may be sold to a producer of an alloy of very similar composition so that adjustment has to be made to the proportions of only one or two elements. A third alternative is to sell the scrap on the basis of the constituent elements of greatest value for the

141

manufacture of lower-grade alloys, but this is wasteful in that some of the alloying elements are effectively lost. The structure of the secondary non-ferrous metals industries varies with the different metals, which are now considered separately.

Copper

The principal consumers of copper-base scrap and residues are:

(i) semi-manufacturers for rolled, drawn or extruded products
(ii) foundries
(iii) secondary alloy ingot melters
(iv) secondary copper refiners
(v) chemical production and miscellaneous users.

Only scrap of the highest quality can be used by the wrought-metal industry. Except for its form (e.g. off-cuts, webbing, bright electrolytic wire, etc.) it is often of similar analytical standard to that of primary metal. Foundries also require good-quality, reliably sorted scrap and normally purchase directly from producers where contamination is unlikely to have occurred. The quality of scrap used by ingot makers is less critical, but it must still be graded according to composition. Selective oxidation by submerged air blowing or flux refining is used to remove unwanted elements, but dilution with primary copper is rarely economic and melting charges must be made up of well-characterized scrap.

Secondary copper refining is much more flexible and a wide variety of copper-base scrap and residues can be present. Although there are differences in the practices employed by individual refiners, the methods used are broadly similar. Copper-bearing wastes are melted in a blast furnace to produce black copper (75–80%Cu). The black copper is then purified by selective oxidation in the presence of a slagging agent, typically in a converter. Higher-grade scrap may also be fed directly to the converter. By careful control of the oxidizing conditions it is possible to produce a copper of 98·5% purity (blister copper). Blister copper is then melted in an anode furnace, deoxidized, and cast into anodes for electro-refining. In recent years the importance of scrap as a feed to the UK refineries has declined and there has been a tendency to work with a higher proportion of imported blister copper.

Copper scrap and residues are used in the manufacture of chemicals, and some secondary copper refining slags containing 1 – 2% copper are used to produce shot-blasting grit. The total copper consumption in these applications is small.

Aluminium

About one-third of the secondary aluminium used is derived from old scrap, the remainder from process scrap and residues. New process scrap arising from foundries and wrought product manufacturers is largely remelted *in situ*. During melting, oxide accumulates on the surface of the bath, from which it is periodically removed as dross. Drosses contain entrapped aluminium at levels of 30–90%, and dross reclamation forms an important part of the secondary aluminium industry.

The amount of refining that the secondary aluminium melter can undertake is severely restricted by the chemical reactivity of aluminium. Of the commonly occurring impurities, only magnesium can be removed by refining, and so the production of secondary aluminium is largely a process of blending and melting. Alloy specifications are met by judicious selection of the scrap input and the addition of virgin metal as required; sorting and pretreatment of scrap is therefore important and a comprehensive scrap classification system exists which defines the form, shape and amount of contamination acceptable in purchased scrap. Until the early 1960s the scrap metal merchants undertook most of the sorting and grading, but rising labour costs and the increasing complexity of aluminium alloys have lead to a deterioration of quality offered by the merchants. The secondary melters now undertake most of their own sorting, making increasing use of mechanical handling methods and new sorting processes.

Zinc

Secondary zinc makes only a minor contribution to UK zinc consumption. Indeed, about half of the recycled zinc is in the form of brass scrap and does not enter the secondary zinc cycle *per se*. Very little zinc is recycled as metal from zinc-bearing wastes. The reason is the difficulty in producing metal of the required specification, which effectively precludes the addition of secondary refined zinc in the production of zinc-base die-casting alloy under British Standard BS 1004. Zinc for galvanizing must contain less than 0.25% aluminium and no copper, and only primary zinc is used. Nearly all secondary zinc is used for the production of zinc oxide, zinc dusts and other zinc chemicals.

Lead

The largest single application of lead is in the lead–acid battery. The high tonnage and short working life of batteries has led to an efficient recovery system which brings about 90% of them back to the secondary smelters. The other source of secondary lead is soft lead and cable alloy (e.g. plumbing scrap), but the use of lead outside of the production of batteries is showing a static or declining consumption rate.

Soft lead is comparatively easy to melt and refine and much of it is merely remelted into ingots. If the lead contains significant quantities of the impurities tin, arsenic, antimony and copper it is necessary to purify the material, for which purpose a refining procedure similar to that used in primary lead manufacture is applied.

Recovery of battery scrap, on the other hand, has acquired its own technology. Batteries arrive at the recovery plant drained of acid and the outer casing is removed either manually or by mechanical crushing. All the lead values are smelted with coke and fluxes in blast or rotary furnaces, in one or two stages. Depending on the processes used, both lead of low antimony content (suitable for making oxide pastes) and antimonial lead (for casting grids, bridges and terminals) can be produced. A recent development is the direct treatment of whole batteries by the Bergsoe process, in which the whole batteries are fed to a shaft furnace fitted with an afterburner to prevent emission of the noxious gases formed from the plastic cases. Combustion of the cases supplements coke in providing energy for the reduction process.

Tin

The largest use of tin is in tinplate and the greatest losses occur in discarded tin cans. Only about 500 tonne/year of tin is recovered from new tinplate scrap by a hydrometallurgical process involving dissolution of the tin coating in hot caustic soda and electro-winning from the stannate solution. Used tin cans are not currently detinned because of the need to find an acceptable method of removing food residues and lacquers, and of opening the soldered seams to allow the leaching reagent to reach all the tin. Detinning is only economic if the detinned steel can be recycled, and in order to reach sufficiently low levels of tin complete access by the detinning solution must be possible.

Apart from the tin used in plating, most recycled tin remains in alloy form. It has been estimated that some 70–75% of reclaimed tin comes from copper- and lead-base alloy, the remainder being from tinplate and tin-base alloys. The most important activity of the secondary tin industry is manufacture of solder, using drosses and fume from the secondary lead and copper industries, residues from solder users, and scrap alloy.

USES OF SCRAP

It is clear that the end-uses of scrap depend upon its quality and that the success of the scrap industry is based upon its ability to collect, identify, grade and process materials

Table 3   Use of ferrous scrap 1973 and 1977

	1973		1977	
	10^3 tonne	%	10^3 tonne	%
In blast furnaces and sinter plants	785·3	4	636·1	4
Steelworks and steel foundries	13 862·2	73	11 222·1	72
Iron foundries	4 098·2	22	3 631·8	23
Refined iron works	204·9	1	194·5	1
totals	18 950·5	100	15 684·5	100

Table 4   Use of non-ferrous scrap 1973 and 1977

		1973		1977	
		10^3 tonne	%	10^3 tonne	%
ALUMINIUM	secondary ingot	177	89	140	82
	direct use	23	11	31	18
COPPER	secondary refined	95	35	76	37
	semi-manufacture	131	48	90	43
	castings and miscellaneous	45	17	41	20
LEAD	secondary refined	145	64	125	72
	scrap lead and remelt	82	36	49	28
ZINC	galvanizing	1	1	2	3
	brass	51	57	36	51
	zinc oxide	14	16	11	16
	die-casting alloy	2	2	2	3
	zinc dust	12	13	9	13
	miscellaneous	10	11	10	14

to meet the user's specification. According to its grade, ferrous scrap is fed back into the various stages of steelmaking, but some low-grade material (notably used tin cans) is only acceptable to the foundries. The secondary copper industry is very flexible. The highest grades of clean segregated scrap can be used directly in semi-manufacturing or for casting and ingot making, while low-grade scrap and residues can be refined into material which is virtually indistinguishable from virgin primary metal. Careful sorting is essential in the secondary aluminium industry where pyrometallurgical refining techniques are ruled out by the chemical reactivity of the metal. The secondary lead industry is domi-

nated by closed-loop recycling of storage batteries, and for this reason the efficiency of lead recovery is apparently higher than that of most other non-ferrous metals (see Table 2). Secondary zinc, on the other hand, mostly goes into the manufacture of zinc chemicals, and the recycling efficiency is low. Recycling of tin from its major end-use in cans is inhibited by technological limitations, and most tin is recycled directly in alloy form or recovered from fume for solder manufacture.

End-use statistics for ferrous metal and for aluminium, copper, lead, and zinc scrap are given in Tables 3 and 4.

U K METAL LOSSES

The problems of upgrading metal-bearing waste to meet end-use specifications are greatest for mixed contaminated wastes and old scrap. Detailed statistical information is lacking in this area but an attempt to bring together the available information was made by Warren Spring Laboratory in 1976. It is important to emphasize that such statistics as are available do not fully reflect the continuously changing situation in industry; they can only be regarded as providing an indication of the size of the problem. Some of the wastes under consideration are marginal materials which can be recovered in existing facilities when demand and prices are high, while others may be exported for processing.

Quantified losses of ferrous material occur in the following major areas:

dusts, fumes and slags from the iron and steel industry;
ferrous metal in domestic and mixed industrial solid wastes (i.e. non-process wastes, packaging, etc); and
the ferrous content of residues from other industries.

These wastes are summarized in Table 5. A further large but unquantified loss occurs as a result of corrosion during use.

Table 5 Main ferrous wastes in UK

Type of waste	Unrecovered Fe, 10^3 tonne
Steel industry fumes, dusts, etc.	
BSC	200
private sector	100
Slags	1 000
Spent pickling acid	20 (at least)
Domestic refuse	1 000
Mixed industrial waste	375
Red mud	35
Titanium dioxide effluent	140
total	2 870

The major areas where loss of non-ferrous metals occurs are:

domestic and mixed industrial waste;
incomplete dismantling of discarded goods;
dispersion of metal during use;
pyrometallurgical residues;
electrochemical refining and metal finishing wastes; and
other chemical wastes.

The annual quantities of metal in these wastes are summarized in Table 6, the data of which

Table 6 Identified non-ferrous metal wastes in UK

Type of waste	tonne/year					
	Cu	Al	Zn	Pb	Sn	Ni
Pyrometallurgical residues	2 700	1 000	35 000	6 000	600	500
Steelworks effluent (Teesside only)	—	—	300	100	—	—
Electrolytic refining and metal finishing	1 000	—	1 600	300	200	1 300
Spent catalysts	200	300	—	—	—	—
Other chemical uses	—	—	13 400*	11 000†	—	—
Galvanizing	—	—	1 500	—	—	—
Mixed solid industrial wastes	10 000	10 000	7 000	3 000	3 000	—
Domestic refuse	24 000	70 000	17 000	8 000	7 000	—
Unrecovered old scrap	41 000	50 000	16 000	—	—	—
totals	78 900	131 300	91 800	28 400	10 800	1 800

*Includes 10 000 t/year in tyres
†Dispersed as antiknock additives

were derived from surveys carried out by Warren Spring Laboratory, the Aluminium Federation, the Zinc Development Association, and published sources. Many of the figures are several years old and it is likely that because of industrial recession some of the current figures would be 10–15% lower. The most notable feature in this table is that a very high proportion of losses of the five major non-ferrous metals occurs in wastes other than those produced in industrial processes. Furthermore, some of the metal-bearing process wastes are recycled or exported for processing—this is true for zinc oxide fume and anode slimes from the secondary copper refiners. Zinc is exceptional in the magnitude of its losses in pyrometallurgical residues, for nearly 7 000 tonne/year is lost in brass melting flue dusts and about 15 000 t/year in steel-making dusts. Metal losses in effluents and sludges are not very significant in quantity, although their environmental impact may be considerable.

The major sources of unrecovered non-ferrous metals are unreclaimed old scrap, domestic refuse, and mixed solid industrial waste. In order to gain a better understanding of why old scrap is not recovered, the Aluminium Federation carried out a detailed study of end-uses of aluminium and used assumptions about the average lifetimes of goods to estimate old scrap arisings. The results are shown in Table 7. The biggest losses of aluminium clearly occur in packaging and domestic equipment. The problem which inhibits non-ferrous metal recovery from domestic and similar mixed wastes is that the metals are present in low concentration and are highly contaminated.

An unknown proportion of unrecovered non-ferrous metal is present in ferrous scrap due to incomplete separation by the scrap processors. Some typical concentrations of non-ferrous metals in ferrous scrap are given

Table 7 Estimates by Aluminium Federation of old scrap arisings

End use	1974 arisings, 10^3 t			1980 anticipated arisings, 10^3 t
	Total	Recovered	Lost	
TRANSPORT				
cars and caravans, Landrovers	20·3	15·1	5·2	30·2
buses, lorries, containers	29·4	19·8	9·6	47·1
marine	1·3	0·5	0·8	4·2
rail	0·7	0·5	0·2	0·9
aircraft	9·0	5·5	3·5	10·1
other	0·6	0·2	0·4	0·4
ELECTRICAL ENGINEERING	4·5	2·1	2·4	8·2
BUILDING, CONSTRUCTION, ETC.	15·2	3·1	12·1	24·5
ENGINEERING AND INDUSTRIAL				
PLANT AND EQUIPMENT	11·8	6·7	5·1	19·8
PACKAGING				
beer barrels	4·2	3·3	0·9	1·6
foil	36·5	0·7	35·8	45·8
other packaging	20·5	—	20·5	38·4
DOMESTIC AND OFFICE EQUIPMENT				
domestic, office and medical	24·1	6·2	17·9	21·6
hollow-ware	5·3	1·6	3·7	9·6
MISCELLANEOUS, INCL. DEFENCE	14·0	4·2	9·8	24·3
*totals**	197·4	69·5	127·9	286·7

*Small difference between total 1974 arisings and Al figure in Table 6 due to rounding error in Table 6

Table 8   Non-ferrous contamination in ferrous scrap

Type of scrap	Cu, %	Sn, %	Ni, %
Basic scrap	0·25	0·025	0·09
No. 4 bales	0·09	0·010	0·09
Pressed and sheared scrap	0·55	0·055	0·18
Fragmented scrap	0·30	0·035	0·25

in Table 8. If the average tramp copper content of steel is assumed to be 0·1–0·2% then the quantity of copper in steel amounts to 20 000 – 40 000 tonne/year. Some of this copper is circulated in new ferrous scrap, but a significant proportion of it must be accounted for in unrecovered old scrap in Table 6.

POTENTIAL FOR IMPROVEMENT IN RECYCLING EFFICIENCY

Recycling of high-grade metal scrap is carried out very efficiently in the UK. Recycling of some of the lower-grade wastes, however, is subject to severe technical and economic constraints; even to maintain the present position, let alone to improve recycling efficiencies, presents a challenge to engineers, metallurgists and chemists. The value of scrap is determined by primary metal prices which in turn depend upon the world supply and demand situation. This means that the fluctuating value of scrap and the cost of processing it to meet a specification are frequently out of step. Since 1973 the value in constant money terms of copper and zinc has fallen steeply and increases in value of other metals have not kept pace with inflation (see Table 9). In the same period recovery costs have risen considerably, e.g. labour costs per unit of output and fuel costs in the metal manufacturing industries have risen by over 160% between 1973 and 1977. Other important costs such as spares, industrial gases, transport, and municipal rates have also increased. The effects are particularly severe on low-grade materials because of the labour-intensive nature of operations such as dismantling and hand sorting. These grades can seldom be used by direct consumers of scrap and frequently must be transported long distances for recovery by secondary refiners.

Leaving aside the question of economic viability, there are a number of areas where there is considerable scope for savings in resource consumption. In the ferrous field the reclamation of used tin cans from domestic refuse offers the greatest potential for increased recovery. Waste sorting plants such as those under construction in Doncaster and Newcastle upon Tyne will provide a ferrous product which is reasonably clean and free from entrained refuse and is suitable for use in ironmaking. The full potential can only be realised by recycling the ferrous and tin contents separately, however, and further research is necessary into methods of cleaning in preparation for detinning used cans.

Recovery of the ferrous content of steel-making dusts depends upon finding a way to remove undesirable elements so that these materials can be treated to yield an acceptable feed for the blast furnace. At the same time there are prospects for recovery of non-ferrous metals, particularly zinc. Two possibilities for zinc recovery arise, which might also be applied to fume from brassmaking: intermediate products may be formed which would be suitable as feedstock for pyrometallurgical zinc facilities in the UK, or a hydrometallurgical and electro-winning process could be developed to treat the wastes directly. One obstacle to zinc recovery is the limited size of the market for zinc dust and zinc chemicals. The processes must therefore provide zinc of equal purity to virgin metal. Both methods are currently being studied, but reference to Table 9 shows that the real price of zinc has fallen drastically since 1973, so that the economics of recovery do not appear to be favourable at the present time. Recovery of non-ferrous metals from domestic wastes presents difficulties because of the wide range

Table 9   Values of non-ferrous metals (constant money terms)

	1973, £/tonne	1977, £(1973)/tonne
Cu	727	407
Al	243	353
Zn	345	183
Pb	174	191
Sn	1 962	3 339

and low concentration of metals present. Nevertheless the separation of products such as waste-derived fuel, ferrous metal and glass generally produces an enriched residual product containing non-magnetic metals such as aluminium, copper and its alloys, stainless steel, die-cast materials and very low concentrations of precious metals. In the USA aluminium is the predominant constituent because of the widespread use of the aluminium beverage can. This has led to the development of technology for the recovery of aluminium from refuse by both wet and dry methods. The most popular approach is the eddy-current separator, basically a linear motor, in which metallic materials are deflected from a conveyor and suitably positioned splitters provide a concentrate of the desired product. Aluminium cans are not present to any great extent in UK refuse, but preliminary work at Warren Spring Laboratory using a ballistic method resulted in the recovery of the copper-rich and aluminium-rich fractions shown in Table 10. It is proposed to investigate the possibility of adding a non-ferrous metal module to the Doncaster plant circuit once the viability of the process core has been established.

An alternative which has been investigated recently is the encouragement of a voluntary collection scheme. Specially printed plastic sacks, showing the Government's 'Save and recycle' symbol and the Aluminium Federation's 'Aluminium is recyclable' symbol, together with a list of acceptable and unacceptable items for collection, were distributed to schools in Buckinghamshire. The weight of scrap collected in a three-month trial period was only 4·3 tonnes, although its bulk was considerable and the quality of the scrap was good. It is felt that voluntary collections should continue to be encouraged.

Better recovery of non-ferrous metals from ferrous scrap is important not only because of the intrinsic value of the non-ferrous metals but also because their presence reduces the value of the ferrous scrap. When a scrap grade contains more of these tramp elements than the steel specification permits, higher-priced iron units must be added to dilute the residual content of the scrap. Conventional fragmentation units can produce good-quality ferrous scrap, but they still rely largely on hand sorting to keep down non-ferrous contamination to acceptable levels and to sort and recover individual non-ferrous metals and alloys. Certain advantages have been claimed for cryogenic fragmentation, although its economic viability remains to be proved. With this technique it is possible to embrittle steel while non-ferrous metals such as aluminium and copper retain their ambient-temperature ductility. After fragmentation, size differences as well as magnetic separation can be exploited to separate non-ferrous from ferrous metals. Separation of non-ferrous metals from each other requires even more sophistication but new developments are being pursued using linear motors, heavy media, and magnetic fluids.

Table 10   Composition of non-ferrous metal products separated from Doncaster refuse by physical processing

Component	Al-rich product, wt-%	Cu-rich product, wt-%
Al alloys	79	3
Brass	4	55
Composites (mainly rubber and brass)	2	4
Copper	1	12
Lead	—	7
Nickel	—	7
Zinc (die-cast)	—	4
Magnetics	4	7
Miscellaneous (combustible) material (mainly rubber)	10	1

Components in each fraction determined by hand sorting

As mentioned previously, the pressures for recovery of metals from metal-finishing wastes are largely environmental. As far as the national economy is concerned recycling of the metal content of these wastes would make very little impact on imports, and arisings at individual factories are generally too small to permit an economical recovery operation. A recent survey carried out in the West Midlands, however, has revealed losses of 100–200 tonne/year each of nickel, copper, and cadmium in about 100 square miles of a heavily industrialized region. The setting up of treatment plants administered by the Waste Disposal Authority is being considered so that savings in disposal costs could be credited to a recovery process. Such centralized treatment plants might be completely self-financing while reducing the extent of pollution of the environment.

FUTURE OF RECYCLING

A great deal of recycling of metals already takes place, its amount regulated by the requirements of industrial economic viability. In general this has meant that recycled metals must compete with virgin metals in properties, but at a somewhat lower price. Both demand and secondary and primary prices often fluctuate quite markedly, and these fluctuations make it difficult to bring new, more efficient technology into use. In 1974 metal prices were relatively high and much effort was expended in developing new processes to recycle low-grade metal-bearing wastes. In 1979 the economics of many of these processes seem much less attractive than in 1974. In the long term, however, as non-renewable resources became scarcer prices must rise in real terms and recycling will assume even greater importance. Nevertheless it must be recognised that recycling is no panacea; even if it is increased, the resource savings are quickly swamped by economic growth, i.e. the recycled material becomes a smaller proportion of the increased use of new material.

A point which is often overlooked is that like all processes, recycling must involve some waste. This is best seen in terms of the energy required in collecting and processing scrap and waste, and of the human resources required to manage and run the recycling industry. It can be shown that when the proportion of stainless steel which is recycled over and over again is taken into account, two-and-a-half times as much energy is used in producing the material as is necessary. Of course, the system gives us a benefit in terms of employment in the recycling industry, but the value of this employment in relation to the energy consumed in the industry is a political and economic question which is not easily answered. It is considerations like these, together with a keener awareness of the need for resource conservation, which have prompted the idea of developing a so-called 'low-waste technology'. This is seen as an evolving phenomenon rather than a revolutionary new technology, the study of which must involve careful consideration of the relationships between the following factors:

1 New methods of manufacture — reductions of materials and energy consumption
2 Use of new materials — new alloys and composites
3 Product durability — design for extended life
4 Design for recycling — automated dismantling
5 Socio-economic impacts — changes in the nature of industry towards manufacture and refurbishing of spare parts, effects on employment and the environment.

Not all of these factors are mutually compatible; for example, the use of some substitutes, particularly when polymers and composites replace metals, may make subsequent recycling more difficult. Nevertheless it can be expected that development of technology along these lines will take place and that recycling will continue to play an important part in ensuring a more rational use of resources.

ROLE OF
H M GOVERNMENT

There are a number of ways in which the Government is acting to encourage recycling of waste materials. First and foremost, it must make sure that there is a continuing and increasing public awareness of the quality of goods made from recycled scrap, and ensure their acceptability to the public. It must have knowledge, through discussions with industry, of the economic and technical limitations to increased recycle of waste. At the same time, the Government can encourage management to take an interest in scrap and waste generated in the production process, so as to ensure not only their segregation but also an understanding of the value of scrap and the value of correct separation and treatment. Industry is also encouraged to make use of the UK Waste Materials Exchange, whereby the Government assists companies to find uses for wastes generated by others.

Development of new technology is also important and support for this is provided through the IRE's, the Requirements Boards, and through such grants as the Product and Process Development Scheme (PPDS). The UK also participates fully in the studies carried out by the European Commission into all relevant forms of waste recycling.

The Government works with local authorities to determine a positive role for them in the collection and segregation of scrap, and supports voluntary groups in the collection and direct sale of scrap to using interests. One area of particular concern in consumer wastes is beverage packaging, where different materials are used for the same purpose (e.g. tinplate or aluminium cans, metal or glass containers). A study of this complex issue has been commissioned so that energy, environmental impact, and resource consumption can be related to different methods of packaging.

G. D. Bashford
Design for energy saving

This digression into the application of metals, it is hoped, does not conflict with the overall content of this conference but may further to some extent an understanding of the user's relationship to the materials supplier. Speaking as a professional engineer who has been in the automobile business for over forty years, I shall take as my theme the design possibilities for automobiles in the short and long terms, with energy conservation in mind.

To remind you of the predicted oil energy prospects, Fig. 1 shows the world situation in terms of supply and demand set against the current growth in demand, which is running at 3-4%. It is obvious that if oil production is due to level off by the end of the century while at the same time the demand for oil

The specific case of the automobile industry is chosen to typify the problems facing designers as to future usage of materials, which has to be related to their total energy requirement. It is believed that the last three decades of the present century will see a steep increase in the selection of aluminium in motor vehicle construction, at the expense of steel, with wider use also of plastics.

The author is Group Chief Engineer, BL Cars Ltd, Solihill, UK.

increases at its present rate of growth, the energy situation will then be such that oil prices will be very high as demand outstrips supply. It is thus essential to identify and evaluate potential energy conservation measures which can be applied in the short term. In the longer term, when natural crude oil becomes scarce and expensive, it will be necessary to find ways of meeting the requirement for energy from other sources.

Given the pressing need to conserve energy, what changes can be made to the automobile to effect this economy? Improved efficiencies as regards engines and transmissions, aerodynamics, rolling resistance, and most of all reduced weight will all contribute to this purpose, but the overriding consideration is the need to achieve energy gains.

So the design engineer has to take into account some new principles relating techniques in materials and manufacture as well as the improvement of efficiencies to the consideration of the total energy balance to gain energy credit. For instance, any reduction in vehicle weight afforded by using lightweight materials must evolve around the primary savings resulting from direct substitution of a lighter material for a heavier one, and also the secondary saving occasioned by the lighter structural loads which will then be carried by the chassis and suspension, and the possibility

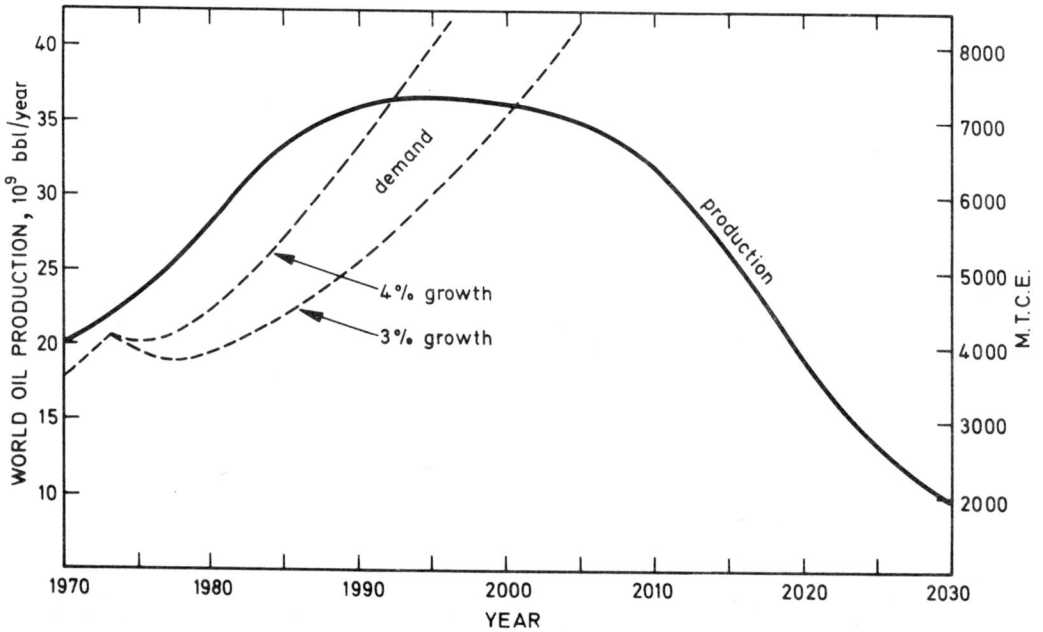

Source: W.R. HAWTHORNE, 'Energy: a renewed challenge to engineers', IMechE Vol. 18952/75

1   World oil supply and demand (MTCE = metric tonnes coal equivalent)

Table 1   Comparison of costs and fuel savings for steel versus aluminium car body structures (based on 50% weight reduction)

Material	Total material weight, kg	Finished body weight, kg	Total material cost, £	Finished body mat'l cost, £	Extra body mat'l cost, £	Potential annual fuel saving at 60% gain, gal	Fuel saving at £1/gal, £	Extra mat'l cost recouped, miles (km)
Steel	302	224	69	64 ⎱				
Aluminium	151	112	177	148 ⎰	84	103	103	9780 (15 740)

Table 2   Car body structures: energy balances for steel versus aluminium (based on 50% weight reduction associated with overall vehicle weight reduction of 30% plus efficiency improvements)

Material	Primary and secondary energy content, kWh/t	Relative energy content factor	Excess energy, kWh/t	Energy balance over 100 000 miles		
				Petrol saving at 60% gain, gal	Energy saving (1 gal = 43 kWh/t)	Energy credit, kWh/t
Steel	6126	1	0 ⎱			
Aluminium	9900 (average)	1·62	3774 ⎰	857	36 851	33 076

of employing lighter-weight power transmission.

In terms of weight saving a wider use of lightweight materials such as aluminium and plastics is fundamental. It is useful to examine this prospect in more detail, keeping in mind that the objective is to save total energy. Taking for example a car body structure and assuming that a 50% weight reduction is achievable on a strength-for-strength basis by using aluminium in place of steel, Table 1 indicates relative values in terms of costs and fuel savings by virtue of reduced weight. At today's prices amortization of the extra material cost (everything else being equal) would be attained in about 10 000 miles.* On the other hand, if there occurs in the near future an escalation in the use of aluminium the relatively high cost of aluminium sheet could well be lowered. There is an apparent abundance of bauxite, but additional smelting and rolling plants would be required to satisfy any large, sustained increase in worldwide demand for aluminium by the transport industry, and therefore a united demonstration of intent would have to be made to the aluminium industry if ultimate weight savings and consequent life-cycle energy gains were to be won by taking this direction. Table 2 fits energy into the picture by showing the difference in the energy requirements of the two materials and exposes the energy credit due to fuel saved over a period of 100 000 miles.

In the context of Professor Alexander's paper on the energy content of materials (*pages 21-29*) it is interesting that in the automotive field, although aluminium is a large consumer of electrical energy—and a short remission of overall power consumption might be expected if the production of aluminium was stopped—the need for energy for transportation would then actually increase. For instance, the energy saved in propelling a lighter car which uses aluminium, over a life cycle of 100 000 miles (160 000 km), could be as much as 29 times the additional energy consumed for the manufacture of aluminium as opposed to iron or steel.

Although this is primarily a meeting for metallurgists, it is important to mention at least briefly the plastics industry, which is

*Based on petrol at £1 per gallon

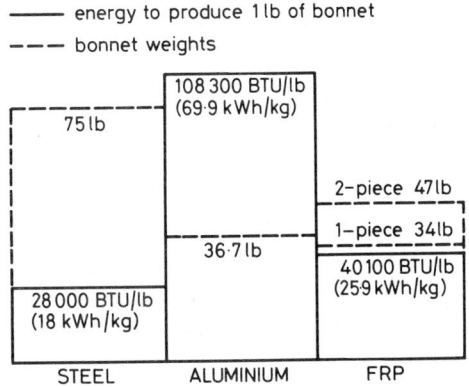

Source: Owens Corning Research 1975

2 Weight/energy relationships for manufacture of car bonnets

without doubt going to play a major role in automobiles in respect of weight-saving. Recent information about reinforced plastics gives a similar picture of the relationship of lightweight materials to energy, as shown in the histogram Fig. 2, which gives the relationships of FRP versus steel and aluminium. FRP needs less energy to produce than either of these metals, while on the other hand aluminium consumes much more energy than steel; so there is a good case for substituting FRP for aluminium provided that all relevant conditions such as impact strength and safety can be met. While this gives a true relationship in energy terms for a single component, it is important to remember that overall component savings have to be related to improved

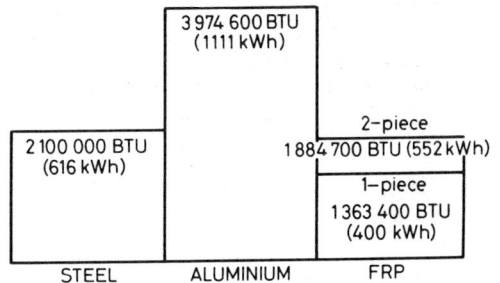

Source: Owens Corning Research 1975

3 Energy requirements for manufacture of car bonnets

fuel consumption; in this analysis I refer to a vehicle 30% lighter. Figure 3 shows the energy relationships for making car bonnets in the three materials concerned.

Whether it be steel, aluminium or plastics, the manufacturer of each material has at heart the interests of his own industry and tends to look at problems from his own committed viewpoint. However, it is generally accepted that materials markets have been disorientated to some degree by the fuel crisis. As the relative values of materials continually shift so their individual status in the industrial scene will also change and their price structures may well modify priorities of use, dependent on energy resources.

The disciplines that designers work to will need to refer back to first principles, and to define functions before selecting materials with a full understanding of their specific properties, which must include their energy requirements. Accurate assessments of overall energy savings which will result from

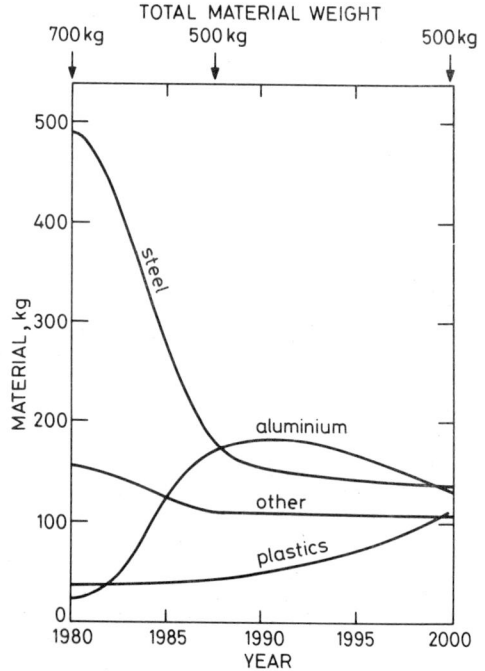

Author's copyright

5   Material weights for future vehicles

changes in vehicle design are difficult to arrive at. They depend on:

(1) reliability of information
(2) accuracy of authors
(3) many unknown factors in manufacture
(4) possible conflict with safety requirements
(5) future projections as to natural, synthetic and renewable resources
(6) size of vehicle and its payload.

There could be endless permutations of future materials, some as substitutions while others become complementary as technology introduces new materials related to price fluctuations. New high-strength low-alloy steels using thinner gauges could be employed for certain applications, particularly where controlled deformability in impact is required, thus providing good safety factors as well as saving weight, yet properly aligned to the total energy concept.

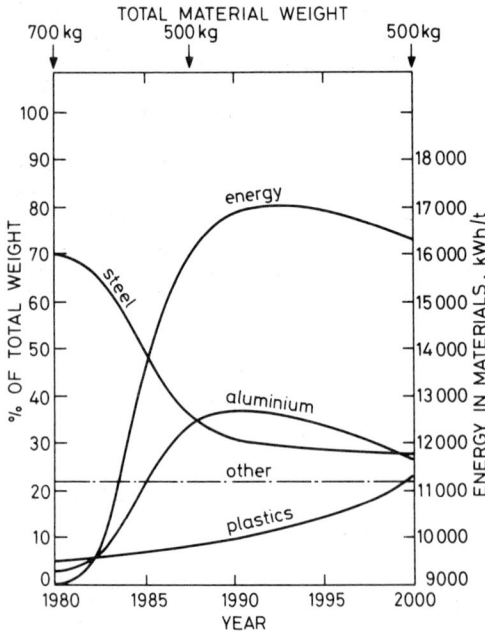

Author's copyright

4   Materials as percentages of total weight of future vehicles

The use of a lightweight material in car construction has a beneficial side-effect, since the reduction of weight makes it possible to scale down engine capacity, and this in turn lowers the amount of exhaust pollutants.

My own company, in planning automobiles for the future, are recognising the need not only to make current models more economical but also to provide for the future substitutions for these cars. They are also aware of what other companies are doing in this respect, such as the public prediction by Volkswagen about the car they aim to market in 1988. In the view of Professor Fiala, Head of Research, the '88 VW will have the following features:

(a) a consumption of 70 mpg
(b) an overall weight lighter by 20%
(c) 30% less air resistance
(d) a turbo-charged Diesel engine
(e) the quietness of a Rolls-Royce.

Predicting the future is naturally difficult, but it becomes surprisingly safe to attempt if one reaches far enough into the future, for the soothsayer can speak out with impunity if he will no longer be around to be confronted with his mistakes; for the more remote the prediction, the greater the probability that an unexpected event will neutralize it. It is accordingly with due humility that I offer my personal observations about the future. The way I see the major materials being used for the construction of automobiles in Europe up to the year 2000 is shown as Figs. 4 and 5. By plotting percentages of total weight, bearing in mind the reduction in overall weight to save running energy, this illustrates a decline in steel and a rapid upsurge of aluminium use, with a gradual climb in that of plastics.

If we now take a long-term view of how the design engineer may be affected in the 21st century, it is quite evident that road transportation will change radically in its method of propulsion and the manner in which people travel for business or pleasure. We may see accomplished the complete relief of city congestion, and one can envisage a fume-free style of mobility employing either an electric city car (Fig. 6) hired at the city perimeter after parking one's personal commuting vehicle, or a system of electrically

powered public service vehicles. There are also likely to be moving pavement systems and extensive underground transport systems, particularly in city areas. The commuting vehicle could be propelled by a coal-fuel, electric, or fuel cell system; there is also a possibility of using hydrogen if the rather intractable problem of its storage can be resolved.

6   Sketch typical of numerous proposals for electrically-powered city runabout

It is logical to assume that design engineers will have to occupy themselves with the present on-going debate over energy conservation, both in manufacturing and running terms, and that this preoccupation will develop into major energy-saving ideas from the mid-'80s onwards. Evolution and science together will gradually define for us the path to be followed; for all we know, the doom-watchers may even have made wrong assumptions, so that we may find we have more fossil fuel reserves than are now believed to exist (although even if this were true there is no escaping the fact that energy conservation will sooner or later become central to our thinking).

While this short paper has confined itself to road transportation, which can only play its limited role in total energy conservation, it is the far greater domestic and industrial demands that must concern us all. Indeed, any form of wastage will have to be avoided and our countermeasures must include the reclamation and recycling which are among the topics at this conference.

W. E. Duckworth
Changing patterns of materials use

It has already been pointed out[1] that the great ages of mankind are named according to the products of the minerals industry: the Stone, the Copper, the Bronze, and the Iron Ages. This has almost certainly given the impression among materials technologists that the dominant engineering material of the 20th century has been steel, although it has been predicted that the production of plastics may soon overtake that of steel in volume terms.[2] Figures 1 and 2, however, put the position in a somewhat different perspective, for it will be seen that both by weight and by volume timber is still the material most consumed by mankind for his domestic purposes and

neither figure takes account of the fact that 46% of all the timber felled is used as fuel within a few miles of the site of felling.[3] It makes one wonder why some historians have not claimed that we are still, after perhaps ten million years on this earth, in the Age of Timber!

From the volume figures for cement consumption it could also be argued that we have not left the Stone Age, and as will be pointed out later, we may be grateful that we have not. Steel, in volume terms, is seen to be a rather poor third but, for reasons to be discussed, may withstand displacement by plastics for the rest of the century and perhaps beyond.

The other major metals, aluminium, copper and zinc, are one or two orders of magnitude further down the production scale. Since any new major class of materials which may be developed in the future must, on the historical evidence of Figs. 1 and 2, take a century to establish itself, then any comparison of changing patterns of materials use of real significance for the next twenty years must be confined to the major groups of timber, cement, steel, and plastics, within each broad classification of which there is, of course, a wide variety of individual material types. Some reference to aluminium, copper, zinc, and the minor engineering materials must be made for completeness.

Aluminium and plastics have emerged as the major new materials of the twentieth century. A relatively stable price pattern between these and the other major materials, timber, cement and steel, has now been established and may only change gradually as energy costs rise. Unless this pattern changes drastically, mutations in the pattern of use depend upon economic opportunity because newer materials are most readily accepted in rising markets for appropriate products.

The author is with the Fulmer Research Institute Ltd, Slough, UK.

156

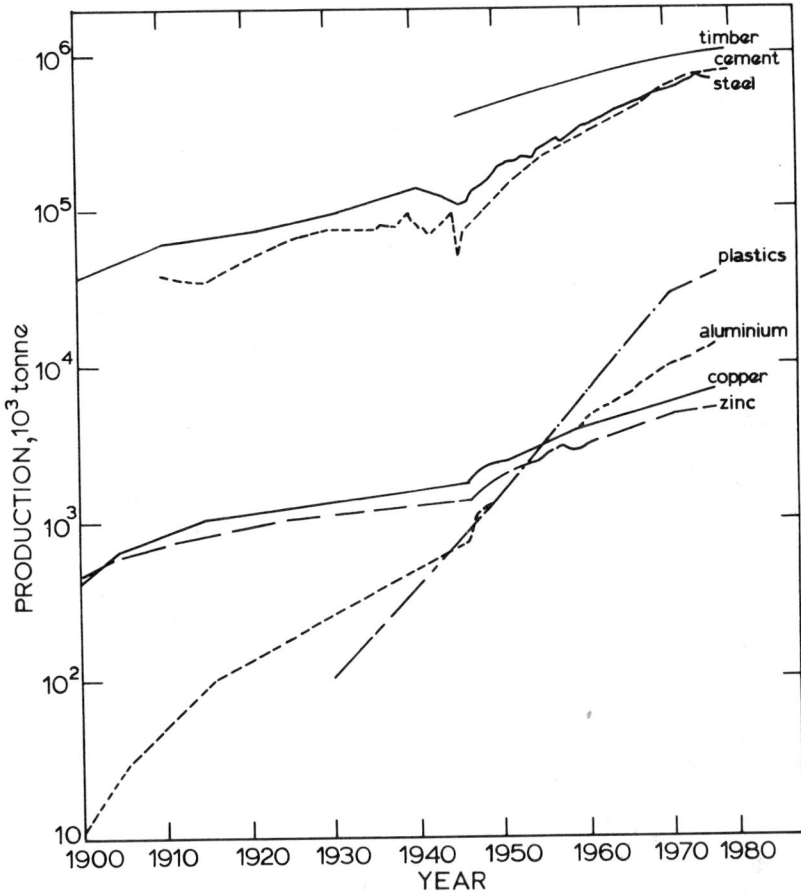

1 Production of materials by mass

Perhaps the most remarkable feature of Figs. 1 and 2 is the apparent resistance of established materials to replacement by the increasing production of newer ones. The volume production of steel was accelerating until the world trade recession of the 1970s, despite the increasing challenge of cement in construction uses and plastics in domestic goods and general engineering. Until the war-time dip from 1940 to 1945 the production of steel also appeared to be serenely undisturbed by the rising output of aluminium. Whether or not this past behaviour is a guide for the future will be examined.

CRITERIA FOR MATERIALS SELECTION

Materials have always been initially selected for their convenience and availability. This is why 92% of the timber produced is used in the country of origin.[3] In the developing countries most prolific in timber it is without doubt the dominant material. It is also relatively cheap.

Selection according to property requirements is the next major criterion and the Stone Age must surely have begun with the realization that sharpened flint was a better

157

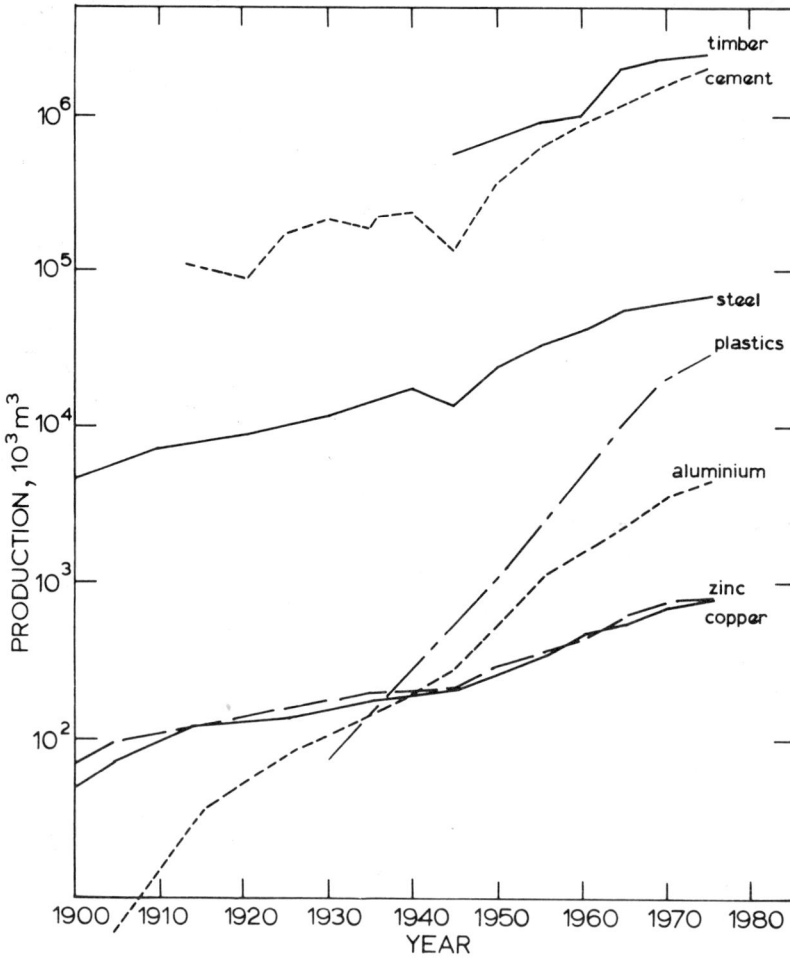

2   Production of materials by volume

tool for skinning animals than even the finest sliver of wood. Even now the derivatives of flint are used for the majority of grinding operations and abrasive uses, such is the conservative disposition of engineers to cling to familiar materials. This is why, for all the extension of technology, the ages of mankind have been characterized by the most convenient and available materials, not necessarily those best for the job.

When the accidental discovery first of copper and then of bronze showed that these were more convenient materials to use for domestic and military purposes, the way was open to the rise of metals technology, and it was only the superior properties of iron for military purposes which led to the development of this metal, initially more expensive than gold[4] because of the high smelting temperatures required. Once the smelting

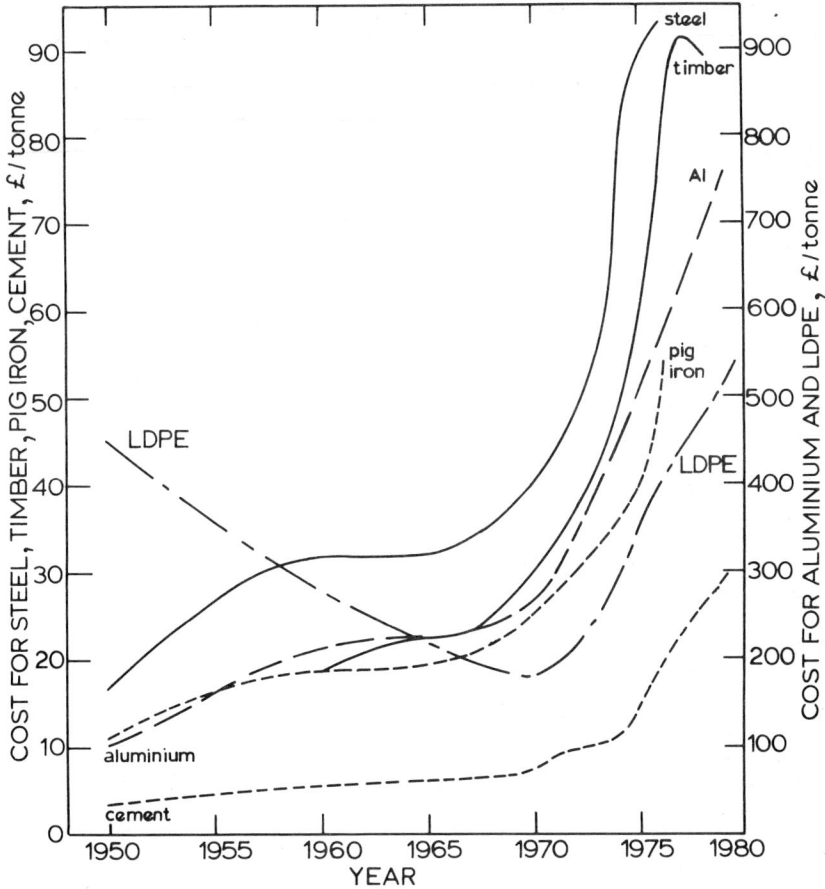

3   Prices of materials in £/tonne

technology was developed, however, its ready availability in conjunction with charcoal and limestone caused its more extensive use and rapidly reducing cost.

Plastics, again, appeared on the scene because of their convenience in use and ease of handling. Their initial cost was high and availability low. Their availability, and hence their falling cost, has been immensely helped by the shift from the initial cellulose-based feedstock to the naphtha by-product of oil refining operations. If oil had not been discovered to be such a convenient fuel, plas-tics might have been a very poor relation to the timber which would have been its feed-stock, along with coal.

Convenience and availability have always been prime factors in materials choice because they have tended to determine the price, which has thus been a consequence of avail-ability rather than a determinant of it. Increasing availability and falling price have thus stimulated further production, this leading to greater availability and still lower price until an equilibrium position is approached and a stable position in the

159

Table 1   Energy content and price of materials

Material	Energy content		Price	
	GJ/tonne	GJ/m^3	£/tonne	£/m^3
Timber	5·2	2·6	180	90
Cement	8	17	29	62
Steel	48	375	148	1 170
Plastics (LDPE)	94	86	525	480
Aluminium	327	844	710	1 830
Copper	61	1 100	537	8 800
Zinc	68	449	400	2 640

materials hierarchy appears to be reached. A comparison of Figs. 2 and 3 shows that the falling price of plastics helped to stimulate a rise in production more rapid than for the older-established materials but that once the price reached a minimum in 1970 it then became subject to the general inflationary pressure, and from that point the rise in plastics production paralleled that of the older materials.

Similarly, in 1960 the ratio of the price of aluminium to that of steel reached its lowest value of around 5:1. It had been around 20:1 in 1900 and is currently about 7:1. From 1960 the growth rate of aluminium production has almost matched that of the older materials. This observation is in line with the findings of Nutting[5] that the relative consumption of metals is dependent upon relative price. As relative price stabilizes so will relative consumption and hence, ultimately, production.

It is possible therefore that we have seen the end of the replacement of steel firstly by aluminium and then by plastics; apart from moderate inroads into specialized areas where properties offer a particular advantage we are probably in a situation where stability of use will continue for the next twenty years. This will depend upon whether the conditions which lead to the remarkable stability of the production relationship between the well-established materials (timber, cement, steel, copper, zinc) will still obtain in the future so that plastics and aluminium will then fit comfortably into their present position within the hierarchy. This will, in turn, depend upon the

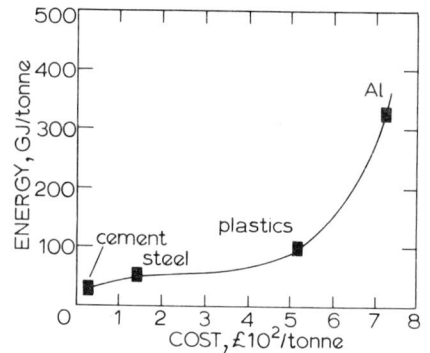

4   Energy content of materials on weight basis

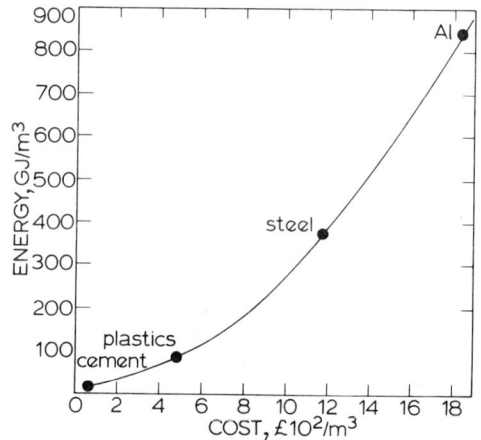

5   Energy content of materials on volume basis

continuing stability of the price relationship and the persistance of conservatism in the criteria for materials selection.

General price levels are difficult to predict, apart from the fact that they will be at least an order of magnitude higher by the end of the century; but relative prices, assuming stability of supply, do appear to depend upon energy content. Table 1 and Figs. 4 and 5 show that there is a strong relationship between the energy required to produce materials and their cost in both weight and volume terms. Copper and zinc are seen to be exceptions because in their cases scarcity

factors begin to affect the position. Timber is, of course, in an unusual position, not being a man-made material.

As energy costs rise (as they are expected to do quite sharply over the next twenty years) timber, cement and plastics will become progressively more attractive in price than steel and aluminium on a volume basis. Scarcity factors are unlikely to affect any of these materials before the end of the century since all have plentiful sources of supply. Even plastics production only consumes 2% of the oil feedstock.

It is unlikely, therefore, that aluminium will do more than hold its own in relative

Table 2   UK steel consumption (10^3 tonnes)

Industry	TOTALS						
	1967	1968	1969	1971	1972	1973	1975
Coal mining	480·8	465·3	443·3	518·8	468·2	515·4	524·7
Food, drink and tobacco	139·3	144·4	150·8	126·7	133·6	143·2	94·9
Chemicals	171·6	175·7	157·4	80·6	92·0	95·0	94·3
Iron and steel	1 141·1	1 387·6	1 488·3	1 019·6	1 139·7	1 387·1	1 109·4
Ironfoundries	83·4	93·8	89·6	83·6	83·6	94·2	71·8
Agricultural machinery	186·2	198·0	208·4	209·8	220·2	261·1	244·5
Machine tools	169·3	190·4	223·0	157·6	155·4	188·3	188·7
Contractors' plant	380·1	393·3	414·4	347·2	341·9	386·2	527·1
Other machinery	722·9	734·2	818·2	700·6	676·9	701·7	631·4
Other non-electrical engineering	536·1	572·4	582·4	543·7	562·6	714·7	513·0
Industrial plant and steelwork	1 823·8	1 773·0	1 760·9	1 671·0	1 484·6	1 610·9	1 461·2
Tools and implements	115·6	124·9	128·4	126·7	136·3	164·6	147·3
Electrical machinery	342·9	349·8	343·2	350·4	330·6	373·5	354·3
Domestic electrical appliances	154·5	183·2	187·3	159·0	157·9	206·2	138·3
Other electrical goods	207·3	235·9	237·9	228·7	274·4	289·8	346·0
Shipbuilding	584·6	596·8	670·6	692·7	718·0	644·8	655·4
Motors	2 002·3	2 202·0	2 343·9	2 303·8	2 213·2	2 428·0	1 653·5
Cycles	77·0	78·2	95·6	76·0	81·5	94·2	60·5
Aircraft	48·4	49·1	48·1	29·2	29·4	34·4	44·7
Private rolling stock	98·2	75·8	60·2	58·5	54·9	50·0	72·8
Bolts and nuts	262·7	285·8	259·7	230·5	209·4	228·5	213·6
Wire and wire manufactures	1 081·3	1 154·7	1 216·0	1 116·6	1 140·0	1 228·2	959·6
Cans and boxes	749·9	810·7	866·7	849·6	854·8	887·3	755·2
Metal furniture	154·3	183·0	166·0	154·9	199·5	169·8	148·3
Metal windows	105·2	114·1	114·1	101·5	119·9	115·1	126·4
Drop forgings	709·9	754·1	863·4	775·7	724·1	857·5	801·1
Hollow-ware	211·6	221·3	234·4	379·0	376·4	463·5	337·7
Other metal industries	1 040·2	1 141·7	1 321·7	1 153·5	1 216·1	1 265·0	863·0
Construction	934·6	989·2	1 004·3	1 456·2	1 498·6	1 737·6	1 070·8
Gas, electricity and water	177·4	159·3	166·6	195·8	202·0	200·3	193·9
Transport	290·1	291·4	318·6	316·5	317·1	317·6	463·5
Other consumers	408·1	441·4	507·9	507·5	537·0	591·7	578·9
TOTAL CONSUMERS	15 590·7	16 570·5	17 491·3	16 721·5	16 749·8	18 445·4	15 445·8

Table 3   UK consumption of primary and secondary aluminium by end-uses (tonnes)

	1966	1967	1968	1969	1970	1971	1972	1973	1974	1975	1976
Transport	123 253	124 414	135 340	147 398	134 000	127 809	131 825	158 190	143 682	117 200	123 700
Mechanical engineering	27 281	25 401	27 431	30 731	28 300	23 946	24 994	36 467	41 588	31 300	34 900
Electrical engineering	59 668	62 648	64 220	60 489	64 600	56 006	56 012	65 378	78 158	66 100	63 100
Building and construction	35 061	34 476	36 353	36 680	36 000	34 416	42 238	54 907	61 979	51 500	59 800
Chemical, food and agricultural appliances	8 033	8 703	10 382	12 425	15 300	13 122	10 841	7 218	7 589	5 700	3 700
Packaging	31 145	30 931	36 732	36 237	35 100	33 954	38 290	47 719	53 991	43 100	102 700
Domestic and office appliances	39 334	39 994	47 675	45 396	43 400	39 947	44 864	55 504	49 020	44 800	50 600
Powder consuming industries	7 860	6 145	7 622	8 005	10 700	7 222	8 374	7 448	11 500	7 600	9 200
Steel industry	15 630	14 010	15 873	19 684	21 700	21 530	23 283	21 397	21 623	18 400	23 100
Metal industries	7 506	8 110	8 060	7 923	7 500	8 068	9 223	15 784 }			
Miscellaneous	48 162	46 947	49 802	57 555	58 800	52 555	63 085	54 421 }	71 627	52 900	71 400
Direct exports of semi-manufactures	50 716	42 392	43 816	42 361	36 600	30 615	35 408	50 670	73 200	59 800	65 200
Totals	453 649	444 171	483 306	504 884	491 600	449 190	488 437	575 103	613 957	498 400	607 400

Source: OECD, as from 1975 EPAA

terms; it may even suffer some decline unless lower-energy techniques of extraction are developed. Steel may suffer in relation to plastics and cement-like materials. To examine the likely trends of changing use if relative prices move against aluminium and steel it is necessary to study in some detail the patterns of existing use of these materials.

FACTORS AFFECTING MAJOR USES OF MATERIALS

Tables 2 and 3 show the relevant figures for steel and aluminium in the UK as a typical industrial country. It will be seen from Table 2 that, in the difficult years 1967–75 total steel consumption grew, albeit unsteadily, and there appeared to be little loss of major markets except in chemicals, private rolling stock, nuts and bolts. Apart from the latter there are special reasons for the decline in each case; the drop in nut and bolt manufacture indicates a change to other fastening techniques such as welding and, perhaps, the use of devices made in aluminium or plastics.

Some markets appear to have successfully withstood the challenge of other materials. The most striking are agricultural machinery, tools and implements, domestic electric appliances and other electrical goods, shipbuilding, motors, hollow-ware and construction.

Table 3 shows that aluminium has made most progress in packaging and in building and construction, with more modest gains in domestic and office appliances, the steel industry, and miscellaneous uses. The modest rise in electrical engineering use is significant in view of the problems with copper during the years in question.

Plastics, as can be seen in Table 4, have made greatest headway in packaging, furniture,

Table 4   USA consumption of plastics (10⁶ lb)

	1969	1970	1971	1975
Furniture	715	826	970	1 420
Automobiles	830	969	1 102	2 080
Packaging	3 343	3 861	4 535	6 490
Domestic appliances	283	330	376	743
Radio and TV	432	503	574	1 002
Building	3 425	2 650	3 987	4 260

building, passenger vehicles, and electrical applications.

The most striking feature of these comparative tables is that all materials seem to expand consumption in the main growth areas. Packaging is the most obvious one, for in this area both aluminium and plastics have shown great growth; yet the consumption of steel in cans and boxes has also grown, although not at the same rate, while Table 5 (timber) confirms that imports of wood for boxboards did not decline. In the same way, the impact of plastics and aluminium in the building industry did not reduce the demand for steel or timber for windows, doors and other construction purposes; and the increasing use of aluminium and plastics in domestic and electrical appliances did not prevent a modest, but uncertain, rise in the use of steel for these purposes. The spectacular rise in the use of steel for hollow-ware until the 1975 recession refutes the impression that there are nowadays only plastics buckets and dustbins available. Similarly, the substantial rise in steel consumption for construction did not affect the pattern of use for cement and timber.

The conclusion drawn from this analysis is that changes in the use of materials occurs most rapidly in a growth situation, most evident in recent years in packaging and domestic appliances. Because more convenient packaging materials were available more packaging was done. Indeed the whole supermarket concept would not have been practicable without plastics and aluminium packaging materials. Steel and timber are not suitable for foil, which has been a major growth area, and yet paper bags are still used in many supermarkets and packaging technology has led to a significant use of steel in ocean-going containers.

There are more plastics used in TV sets and radios because there are more TV sets and radios for which steel would not be suitable and wood is less convenient; but plastics have not replaced wood in the slower-growing furniture industry except in a distinctly down-market area and in the thriving areas of modern kitchen designs.

In historical retrospect this pattern is seen to have obtained for many decades. Steel ships replaced wooden ships because vastly more were built. GRP will not compete with

Table 5 UK imports of wood products in 1966 and 1976 (value, £10³ c.i.f.)

Type of wood or wood product	Value, £10³ c.i.f.	
	1966	1976
Fuel wood	n.a.	135
Wood charcoal	n.a.	684
Pinewood	n.a.	785
Pitprops	3 807	2 031
Telegraph poles	1 304	760
Sawlogs, veneer logs	8 005	16 982
Rough shaped woods	n.a.	4 529
Boxboards	1 002	2 627
Pencil slats	n.a.	949
Planks > 5 mm thick	173 308	527 462
Sleepers	2 046	2 622
Wool and flour	n.a.	464
PTG boards		25 808
Flooring blocks, strips		1 752
Window frames		376
Doors	39 739	7 095
Skirting boards		3 015
Shuttering		59
Sectional building		4 128
Builders' joinery		3 069
	262 561	605 332

steel in a stagnant industry but will compete with wood in the rapidly expanding pleasure craft market. Steel replaced wood in land transport because of the invention of the steam locomotive and the motorcar; on the other hand, aluminium and GRP have challenged too late in an industry which has lost its dynamism and the new dual-phase steels will stifle the opportunity which legislation on fuel consumption might have provided. Concrete and steel have supplemented wood in construction and building because the timber supply would have been totally inadequate for the demand and unsuitable for many applications. As the consumption of cement declined in the UK (Table 6), because of the long recession in the construction industry the pattern of use has remained very stable.

The lesson of the recent and distant past is, therefore, that material substitution occurs most readily in a rising market. The reasons are not difficult to see. Material substitution

Table 6   UK consumption of cement (10^3 tonnes)

	1973	1974	1975	1976	1977
Ready-mixed concrete	794	748	710	631	615
Concrete products	386	340	332	343	308
Asbestos cement	405	402	306	321	302
Merchants' yards	247	231	216	211	217
Site use	456	384	388	333	262
Totals	1 923	1 744	1 676	1 550	1 430

is always risky and if the market is not expanding for the product concerned the risk is less attractive and the likelihood of recovering the necessary development and start-up costs is less certain. The suppliers of a new material will only invest in production facilities if they see a rising market. Where a new material is challenging an existing one in a stagnant situation the opposition of the suppliers of the existing material will be more virulent than if there is clearly room for both materials.

The most recent, and severe, example of the risks of introducing a new material has been Corfam, a 'breathing' synthetic leather developed at a cost of $100 million by Du Pont as a less costly but equally comfortable and durable replacement for leather in ladies' shoes. Unfortunately, it was discovered too late that women replace their shoes because they go out of fashion and not because they wear out. Cheaper synthetics such as vinyls had captured the growth market and could not be replaced by a higher priced product. In a recently published paper Cannon[6] illustrates how a programme for making a substitute material may have to be terminated at a late stage because the commercial risks are too great.

The material world abounds with instances in which a challenge has been fought off, or held back, when made in a stagnant situation: double-reduced tinplate, one-coat enamelled steel, plastic-coated steel, thin-walled die castings, and latterly dual-phase steels are all examples. Glass-reinforced concrete and other developments will help cement to retain its place in the construction market.

The battle between alloy die castings and injection-moulded thermoplastics is a typical example of the factors inhibiting change of

material use unless the market is buoyant. Brandes and Waterman[7] have shown how important energy factors are. In order to recover the investment in research, design, capital equipment, production, marketing, and servicing the product in a new material within a reasonable time the volume of output must be high.

Where market leaders have committed heavy investment in particular materials, as is the case with copper-base alloy fitments for domestic bathroom items in the UK, again there has to be a very rapidly expanding market to encourage competing firms to introduce alternative materials. Legislation can, of course, suddenly create such large market opportunities. In the USA the Nader-inspired safety regulations for automobiles have led to the replacement of chromium-plated steel for bumpers by reaction-injected moulded polyurethane.

FUTURE PATTERN OF MATERIALS USE

From the above analysis it can be seen that

(1) no significant new materials will emerge this century

(2) the existing pattern of materials use can change by the end of the century if
   (a) rising costs of energy cause a significant change in the present price relativities of existing materials

164

(b) technical developments such as a large reduction in the energy cost of producing aluminium cause a substantial change in the price relativities

(c) the world does not experience a twenty-year prolongation of the 'stagflation' caused by the 1973 dramatic oil price rise

(d) major new uses arise for existing materials in which the price/property pattern is most favourable to one or two materials

(e) legislation forces substantial changes.

In attempting to predict the future (a) and (b) should be taken together, as should (c) and (d). Legislation which forces too many uneconomic changes is likely to be successfully resisted on a worldwide basis and so will not make a major impact.

It seems almost certain, as indicated earlier, that rising energy costs will penalize steel and aluminium in relation to timber, cement, and plastics. Since it is unlikely that a revolution in aluminium technology will reduce its price, aluminium is unlikely to find major new markets and may well find some of its

established ones subject to erosion, especially in the most price-sensitive areas. New alloy development to produce better properties on a volume basis could help, but the industry has a poor record of alloy development and such programmes take time to mature.

Steel, being a much more versatile material, is better equipped to match price increases by property improvement and volume reduction, but it must become more conscious of value-added and less preoccupied by tons of output.

Plastics will continue to expand into packaging, transport, and domestic appliances but will not make significant inroads into the steel and timber markets unless (c) and (d) combine to improve the prospects for materials consumption generally. This is because, as Table 7 shows, plastics are more expensive than steel and timber in property terms and the relative price pattern can only change adversely in relation to steel which, thanks to its versatility, should successfully withstand the challenge.

Timber and cement will keep their markets and expand in accordance with their historical trend.

Table 7   Price per unit tensile strength for various alloys and materials in 1969

	Tensile strength σ, 10^7 Pa	Density ρ, 10^3 kgm^{-3}	Specific price, £ kg^{-1}	Price density $p\rho$, pm^{-3}	$A = p\rho/\sigma$, 10^{-1} p N^{-1} m^{-1}
Steel castings	37–124	7·8	0·14–0·22	1 002–1 716	0·015–0·03
Steel weldments	30	7·8	0·26–0·19	2 028–6 162	0·05–0·16
Grey cast-iron	15–22	7·3	0·01–0·12	511– 876	0·03–0·04
High-conductivity copper rod	23–31	8·93	0·60	5 358	0·17–0·23
Brass rod	39	8·36	0·43	3 595	0·09
Aluminium sheet	29	2·7	0·47	1 269	0·04
Magnesium alloy diecastings	19	1·7	0·42	714	0·04
Zinc alloy diecastings	28	6·6	0·21–0·26	1 386–1 716	0·05–0·06
Titanium alloy sheet	96	4·42	11·50	50 830	0·53
Polythene mouldings	1·5	0·9	0·26	234	0·16
Polyvinyl chloride (pipe)	6·2	1·4	0·20	280	0·05
Nylon/glass-filled mouldings	15	1·38	0·93–1·13	1 283–1 559	0·09–0·10
Resin/glass laminates with glass weave and polyester resin (laid)	39–108	1·87	3·10	5 797	0·1–0·2
Concrete					
reinforced	1·5	2·4	0·01	24	0·015
prestressed	2·3	2·4	0·02	48	0·02
Timber	0·8	0·5–0·6	0·03–0·04	15–24	0·02–0·03

Numbers in last column are prices (in p per metre) that will just support a mass of 1·02 tonne in gravitational field of Earth; p = penny (£0·01)

Table 8   Trend of materials consumption 1950–75
in (A) developed and (B) developing countries (10^3 tonnes)

	1950		1960		1975	
	A	B	A	B	A	B
Cement	95	8			325	65
Steel	275	11			525	107
Plastics	1·4	0·1	5·8	0·9	65	10

If world economic growth continues the trend of the 1960s then, if the developing countries mainly benefit, plastics will rise most in consumption. If the developed countries benefit most then so will cement, steel, and plastics in that order.

It can be seen in Table 8 that, for the major materials, consumption has risen faster in the developing countries than in the developed (timber is not included because its major use is in the developing countries as fuel). In the developed countries the rise in consumption has been greatest in plastics and least in steel. In the developing countries it has been greatest, again, in plastics and least in cement.

Table 9 shows a prediction of materials use to the year 2000 which, produced in 1969, was based on production data up to 1966, on the assumption that the pattern of growth of metals and plastics would remain unchanged and that the world population would be 7000 million in A.D. 2000. If China, with one-quarter of the world's population and one-tenth of the materials consumption of the USA, is to set the pace for the developing countries then, since it places greater emphasis on capital than on consumer goods, plastics consumption worldwide is unlikely to have overtaken steel in volume terms by the end of the century. The future for a substantial further rise in the consumption of plastics will thus depend upon a more buoyant consumer goods market, where its outlets mainly lie, in both the developed and developing countries; but the present indication of world economic trends does not encourage the prospects for such a consumer boom. In these circumstances plastics will have a much harder fight to make

Table 9   Production of certain materials
extrapolated to A.D.2000

Material	1966 million tonnes	1966 litres per person	1985 million tonnes	1985 litres per person	2000 million tonnes	2000 litres per person
Iron	464	18	1 130	29	2 250	41
Aluminium	7·7	0·6	55	4	250	13
Total metals including copper and zinc	486	19	1 204	33	2 535	55
Plastics	16	4·2	240	41	1 700	212
Rubbers	3·9	1·2	16	3·4	44	6·6
Total	19·9	5·4	256	44·4	1 744	218·6

further inroads into the traditional markets for steel, aluminium, copper, zinc, and even timber than they have in the past.

Under greatest threat are probably aluminium and copper because of price and political considerations. As Nutting[5] has predicted, increasing price and scarcity will cause every metal except steel to settle into the role which only it can fulfil.

Looking, for a moment, beyond the year 2000 the low-energy indigenous materials, sand, cement, and ceramic materials, will surely find an increasing role as Kelly[8] has pointed out. This is why we may be glad we never really lost the skills of the Stone Age!

CONCLUSIONS

1 No important new materials will emerge in the next twenty years. Significant further changes in the usage of the major common materials — timber, cement, steel, plastics, aluminium, copper, zinc — will depend upon economic opportunity. A stagnating world economy will maintain the current pattern of use except where property and convenience advantages still remain to be exploited, as in packaging and other consumer goods applications.

2 In a more expansive world economy benefiting the developed countries, plastics could overtake steel in volume consumption by 1985 because of rising consumer markets. A world economy led by increasing capital expenditure in developing countries would help cement and steel maintain their current relative positions. Rapidly rising consumer expenditure in the developing countries would, again, be to the advantage of plastics.

3 Aluminium, copper, zinc and the other metals are unlikely to find major new markets.

ACKNOWLEDGMENT

While the statistical information contained in this paper is less complete and comprehensive than I would have wished (it has proved extraordinarily difficult to obtain for certain materials) such information as I have been able to gather has been with the able and energetic help of the Fulmer Librarian, Mr R. F. Flint, and other colleagues, whose assistance is gratefully acknowledged.

REFERENCES

1 M.F. DOWDING: 'The world of metals,' *Metals and Materials*, 1978, July, 27

2 *Plastics and Rubber Weekly*, 17 November 1978, 27

3 'International book of wood,' p.226; 1976, London, Mitchell Beazley

4 L. AICHISON: 'A history of metals', vol. I, 307, 1960, London, MacDonald and Evans

5 J. NUTTING: 'Metals as materials,' *Metals and Materials*, 1977, July/August, 30

6 C. G. CANNON: *Int. J. Materials in Engineering Applications*, 1979, 1, (5), 250

7 E. A. BRANDES AND N. A. WATERMAN: *ibid.*, 260

8 A. KELLY: *Int. J. Materials in Engineering Applications*, 1979, 1, (3), 132

Discussion

Discussion
SESSION A: General

Chairman: Professor J. Nutting

P. C. F. Crowson
Geography and political economy of metal supplies

Mr Y. KOVACH
(Charter Consolidated Ltd):
Would Mr Crowson tell the Conference briefly what Germany and France are doing in regard to stockpiling?

Mr CROWSON:
In Germany there are mixed attitudes. For the 'bread-and-butter' metals, there is a very strong attitude that the German Government should not be involved in stockpiling, so that copper, for example, would not figure in German stockpiling proposals. On the other hand, it has decided to adopt a stockpiling facility of critical raw materials. The debate is still continuing but, broadly speaking, it has become Government policy in recent weeks.

As far as France is concerned, policies are much less publicly debated in the raw material field and it is very difficult to get a clear view of exactly what is happening.

Sir RONALD PRAIN
(Selection Trust Ltd):
I think I heard Mr Crowson say that it would be logical policy for this country to buy from the cheaper producers of copper. Of course, the price of copper that you pay is in no way related in the short term to costs of production; countries like Zambia and Zaire with high costs have to sell at the same price as countries with low costs. Mr Crowson may have meant that it would be in our interest to buy from the low-cost countries because they would be least likely to go out of business in the long run; but I would like to comment that with countries like Zambia and Zaire, however high their costs go they are going to stay in the copper business. The fact is that these nationalized industries are not going out of business, so that I am not sure that it is valid to say that it would be in British interests to buy preferably from low-cost producers.

Mr CROWSON:
I agree, but what I said in fact was that we should purchase from the lower-cost areas while seeking means of reducing any external political obstacles to efficiency in the higher-cost areas (basically Zaire and Zambia) and looking for ways to improve efficiency in the copper belt.

171

Dr W. MACFARLANE
(Wellman Incandescent Ltd):
I would like to ask Mr Crowson about the date of the information in Table 11, which may have had some effect on the final metal cost. Has he any indication of the percentage of metal costs relating to the energy consumed being a factor? — because with the recent increases in the prices of oil and natural gas, the policy of decontrol of prices in the USA and the somewhat different uses of the production of gas from coal in South Africa could lead to significant changes in Table 11. The dating of the information therefore is rather important.

Mr CROWSON:
The date is given under Table 12. The data refer to 1977, the last three columns. As you rightly say, the costs and indeed the grades (average head grades) mined in any one year change dramatically from year to year, but oil and energy costs are not the only factors. Other factors which are more important, especially over the last ten years, have been changes in exchange rates which affect the ranking of each country; broad cost inflation; and the structure of the mining industry in any of these countries from one year to the next. While one should look at energy costs (I do not know the percentage of cost they represent), energy is only one of a number of factors in play.

THE CHAIRMAN:
I wonder if you have any views on the role of China as a supplier of non-ferrous metals?

Mr CROWSON:
China has, reportedly, substantial reserves of some of the metals. It claims large reserves of copper and of all other metals it needs except perhaps chromium; in the latter case, they are on the borders, inaccessible and highly expensive. The cost and timescale of developing those metals beyond what China needs itself tend to be underestimated in the West. China is a major importer at the moment of copper, iron, steel, lead, zinc and nickel. It will be many, many years before China (if ever) becomes a dominant supplier to the West. One of the interesting things as regards its existing copper deposits is that they tend to be very small-scale and very difficult to exploit. Thus, China's costs for its existing production would be something like 500 as against an average of 100 in the Western world.

Mr J. LILLY
(Falconbridge Europe SA, Brussels):
Could you comment on the use of undersea nodules, undersea mining problems, and how politics will affect the sharing of the undersea nodule mining? What are your views on when that material will come on stream?

Mr CROWSON:
I suggest that Alan Archer is far better qualified to comment than I am.*

Dr M. E. HENSTOCK
(Nottingham University):
I would like to ask a question about tin. I am sure Dr Robins is going to cross swords with me about this, but it is my impression that the price of tin bears little relation to the production costs, because these vary widely throughout the world. Would Mr Crowson like to comment upon the discrepancies between mining costs in South-east Asia and South America, and how these are wrapped up into the internationally determined price of the metal?

Mr CROWSON:
I think Sir Ronald Prain has brought this out already, that the mining costs of any metal often bear little relationship to the price of that metal, except in periods of weak demand, as we have recently seen, for example, with copper; but by and large the costs of production of all metals cover a very wide spectrum not only between countries — between, say, South-east Asia in the case of tin, and Bolivia, but also within countries such as Malaysia. The costs of tin mining in Malaysia, if one looks at individual mines (even if one takes the same type of operation, such as a dredging operation), covers a very wide spectrum. In the case of tin, Malaysia has very high taxes and export duties on tin which broadly speaking means that the costs of production including taxes are not markedly different from those costs of production even including taxes in Bolivia. The Malaysians and the

---

*See Mr Archer's contribution to Session B discussion following Professor Nutting's paper on cobalt and chromium *[Ed.]*

Bolivians and most of the South-east Asians regard tax as a cost from the point of view of the nation, whereas a miner might regard it not so much as an operating cost but as a surcharge on top of his costs. I do not think one can really take any metal and say there is at any time a fixed relationship between the price and the costs of production in the very broad spectrum of the mines. At the margin there is, but the margin in the case of tin would not be South-east Asia.

Dr A.F. BONFIGLIOLI
(University of Sussex):
I would like to refer to your chart about the share of less-developed countries. You quote there about 75% of bauxite. Does this refer specifically to raw bauxite or does it also include alumina?

Mr CROWSON
I think it probably includes alumina.

W. O. Alexander
Total energy content of some metals and materials and their properties

---

Dr D. S. FLETT
(Warren Spring Laboratory):
I think Professor Alexander has perhaps dismissed hydrometallurgy in a rather cavalier fashion. Hydrometallurgy has a known history going back more than 300 years and some metals can be won only by hydrometallurgical extraction. I would not argue that a significant swing towards hydrometallurgy seems unlikely for those metals for which, over time, it will still be possible to produce concentrates acceptable for conventional pyrometallurgical processes at an acceptable cost. It is recognised that energy costs for pyrometallurgical processes are lower than for hydrometallurgical alternatives, although capital costs show little difference these days. However, the adoption of hydrometallurgical processes continues to increase. Why is this? In the first place capital costs of hydrometallurgical processes are less sensitive to scale factors and thus can appeal to small operators. More importantly, however, as ore grades drop the cost of mining goes up and in particular the cost of comminution, a very energy-intensive process, will increase and a point will be reached where physical beneficiation methods can no longer produce concentrates for smelting at acceptable costs. At this point hydrometallurgical processing becomes much more attractive with the application of such low-cost processes as heap leaching and *in-situ* leaching. Similarly processing of complex, difficult-to-beneficiate ore bodies may well be more amenable to hydrometallurgical processing. Indeed this trend is already with us in the treatment of low-grade oxide cappings of porphyry copper deposits, bacterial leaching of low-grade sulphide ore, and hydrometallurgical processing of lateritic ores. There is little doubt in my mind that hydrometallurgy has a significant and increasing role to play in increasing the resource base for many important metals and must not be ignored as a major processing route in extractive metallurgy.

Professor ALEXANDER:
Fortunately, I was quoting Professor Kellogg who, I would remind you, is a great diagnostician about the future. I heard a lecture of his 25 years ago when he forecast titanium at 1 dollar a pound. Everybody howled him down but within five years it was very near 1 dollar a pound. However, what worries me about hydrometallurgy (and I am not an extraction metallurgist) is the sheer bulk and extraordinary time cycles which will be involved in the processes you mention. Furthermore, as many of the audience know, I am an anti-powder man. I cannot see that powder metals can ever achieve tonnage significance. They have been fighting very hard for 50 years yet in this country even the tonnage of steel produced is only 15 000 tons a year in the powder form. It has been going on for a very long time, so metallurgically, I think, it is not attractive to

the engineer. However, that is not to say that as an extraction route it is ruled out with remelt facilities and so on.

Dr R. B. NICHOLSON
(Inco Europe Ltd):
I would like to comment on Professor Alexander's remarks about the use of old scrap, when he highlighted the apparently good performance of lead. I think I am right in saying that he was quoting figures in which he showed the percentage of old scrap at current year's consumption. It gives enormous advantage to materials which have very short lifetime and the reason why lead comes out so well is that a lead car battery last for such a short time.

Professor ALEXANDER:
That is an interesting observation but I doubt its validity overall. I have studied the consumption and recycling of copper over 50 years in the UK. One point which needs a comprehensive answer is the average life-cycles and deviations of the main tonnage usages. We all assume the figure which the Americans gave for steel; they said it was 20 years and everybody assumed that, because it applied to capital goods, capital structures, or what the steel industry calls capital scrap, it applies to everything else. I know one can argue that if you scrap the steel part of the structure all the other metal parts are scrapped. But ignoring the actual lifecycles, if you take both Warren Springs' and our own studies of recycling of old copper scrap in the UK, the plain fact is that there is not the old scrap coming back into circulation. One must note also that the consumption of copper in this country has been constant now for 15 years, so by now we should be getting a significant amount back. This is another area where we must make very intensive studies to see where all the old metal has gone after 30–50 years' service.

Dr H. ZIJLSTRA
(Philips Research Laboratories, Eindhoven):
Professor Alexander made a general statement about the comparison of the energy required to make a product and the energy that is going to be saved during the operation of the product. That is perfectly true if the sole object of the product is to save energy or produce energy but I do not think that generally there are many products that are made to save energy. For example, carnations take twice as much energy to raise as roses but I still take carnations for my wife because she prefers them. The Professor was talking about total energy. The total energy required to make a metal or any material can be separated into two parts: one is, in the case of metals, to reduce the metal oxide as we find it from the ore into the metal itself (that is the energy we have to spend according to the first law of thermodynamics); the other part of the energy is to concentrate the ore, and that is the price we pay in the second law of thermodynamics — concentration. The ratio of those two energies may differ very widely for several materials. For instance, in copper or nickel, which are very easy to reduce, when we go to poorer and poorer ores we have to spend more and more energy in the second law, but aluminium and titanium take the greatest part of the energy in the first law, and that energy can be calculated. Any company that is working effectively will reduce that to the minimum set by that law, but it is far more uncertain in the concentration, the second law. If the first law takes the greater part, then we have a very rigid price which will go up and down with the energy price; that helps for titanium and aluminium, but for other materials such as nickel it may vary much more widely. Would you comment on that?

Professor ALEXANDER:
This is why I think it is important to have agreed values of total energy. I do not deny what you are asserting but it comes to exactly what Mr Crowson was saying; that the costing of the energy varies due to one cause and another, and I feel that until we have absolute values of total energy this variation in production or in selling price will persist. As scientists, we must concentrate on absolute values so that in the end we can say 'This is the true situation energy-wise and material-wise'. The ultimate consideration is what properties you are getting in energy terms.

THE CHAIRMAN:
You raised a very important point in relation to satisfying wives' needs. Perhaps the problem is that we shall have to educate our wives to prefer roses to carnations. If we translate these metaphors into industrial

terms, we shall have to think of using materials which are less energy-intensive and designs which are also not energy-intensive. Of course, if we are very clever we might persuade our wives to prefer dandelions and daisies to roses!

Professor ALEXANDER:
If you are growing the roses and carnations in your own garden, using solar heat which is renewable energy, it is not an effective analogy.

THE CHAIRMAN:
The content and style of our next paper was hinted at by Sir Alan Cottrell. In the USA, as we have seen, there are many lists of the ranking order of metals in relation to defence needs. These lists differ considerably according to who has prepared them but at least there are public lists. In this country we never seem to see a list of anything that might relate to a ranking order for *our* defence needs, even though obviously there *are* defence needs and defence may claim a certain priority in relation to its demands for raw material resources. In this context, Dr Dyker from the University of Sussex is going to tell us something about distortion of economic factors due to defence strategy.

D. A. Dyker
Distortion of economic factors due to defence strategy

Mr M. H. DAVIES
(Copper Development Association):
If we take the medical analogy, there has been a great deal of debate and controversy about the disproportionate application of resources to heart transplants. In the defence field, the question I would certainly like to have answered, as perhaps would most of the people here today, is: to what extent have defence requirements distorted the metals picture in the UK and elsewhere? In other words, how much of the total effort is devoted to the development of sophisticated materials and systems compared with what Sir Alan asked for in his Address — an intensification of effort to develop the maximum properties in the readily available materials.

Dr DYKER:
I think what is being asked for is statistics on precisely what the resource flows, particularly in terms of metals, are to defence proper and defence-related activities. I know how one would go about trying to discover this — through input/output tables which give a total matrix of all the interrelationships in the economy. In the case of the Soviet Union, of course, the input/output tables are not systematically published but they have been reconstructed by Western economists. Unfortunately, you just do not find defence present; what is officially reported as defence in the Soviet Union really does not take one very far. In relation to the United States, I would have thought that it ought to be possible to trace all defence-related resource flows. Input/output tables have been constructed for the United States, but I have not seen any systematic quantitative research done on this. I completely agree that it is the '64 000 dollar question'.

THE CHAIRMAN:

It is perhaps interesting that if we look at the attendance here, there are few people present from what might be called the defence sector. Would anyone from the defence sector like to comment, or are they just here to listen?

Mr P. J. KIDDLE
(Climax Molybdenum Co. Ltd):

As you are a Russian specialist, I wonder whether there are data available on the raw material resources available in Russia. Generally speaking, we have very sketchy data on the raw material actually available and their production within Russia. If this sort of data is available in any more detail than is normally published, I am sure it would be of great interest to many of the audience.

Dr DYKER:

This is a very interesting subject and, of course, a good deal of work has been done in recent years on mineral resources in the Soviet Union, in particular a study done by a consultant for the British Ministry of Defence on Soviet gold production; I think we can now say we have fairly firm estimates on that. The problem arises when you come to talk about reserves. A few figures spring to mind: for example, I think that almost 50% of total world coal reserves are in Eastern Siberia, yet the Russians do not systematically exploit these resources because they are all under permafrost. There can be absolutely no doubt that Siberia represents a unique treasure house of energy and metal resources, but most of it is north of the permafrost line while most of northern Siberia is permanently flooded during the summer months for geographical reasons I will not go into. Even the Soviet Government, who are ideologically inclined to be over-optimistic about the extent to which man can dominate Nature, have been extremely pragmatic in their actual policies towards the development of these resources. They start with the cheapest and at the moment, as you are probably aware, they are concentrating on oil and increasingly gas in Western Siberia, although in capital terms these are enormously expensive (there are some published figures on this). Presumably, as they move eastwards so the technological problems will become greater and capital costs will soar. But if your scenario is a world

in which energy and metal resources are becoming well-nigh exhausted, then there may well be a feeling that we must have what is in Siberia at any cost, and then the Russians will have everybody over a barrel. Speaking as an economist, in terms of price relativities or any kind of reasonable assumptions about how prices will move over the next decade or so, I certainly find it difficult to imagine how anything more than a small proportion of Siberia's resources will ever become economic resources. A propos of gold, I recently had a discussion at another conference about just how important it is that the Russians have all this gold. For example, when they have to buy grain after a harvest failure, can they simply pay for it in gold if they cannot get Western credits? The answer is a very simple 'yes' but it might cost them more to get the gold out of the ground than to improve the efficiency of their agriculture.

THE CHAIRMAN:

We will conclude with that significant comment, because I think this may well interact with many of the other things we are discussing in relation to metal use. It has become very apparent that there is a need for a metals forum and for discussion of these defence topics, because it is quite clear that we know very little about this, or very little that can be made public — and some of it does deserve a wider audience and merits more critical discussion.

A. A. Nijkerk
Economics of metals in view of their scarcity

Dr J. BARFORD
(CEGB, Leatherhead):
Professor Alexander and other speakers drew our attention to the critical importance of considering energy requirements in any future metal strategy, and Mr Nijkerk has emphasized the importance of recycling. It is with some surprise, therefore, that I notice the omission of a particular element, which by several criteria is a metal, from the programme of this meeting. I refer, of course, to uranium which I would have though would have been worthy of consideration.

Secondly, the question was raised after Dr Dyker's paper as to what has been the real effect of the defence industries on our materials strategy. I have a feeling that there is a fundamental effect and it seems to me to point to a rather basic law which is almost Cottrellian in its fundamentalness. It goes something like this: 'the amount of development or research that goes into a material is in inverse proportion to the amount sold or used.' I think this is fairly obvious, not only in the metallic field but in the materials field as a whole. Dr Duckworth's paper this afternoon points to the use of concrete, which probably has had the least investigation into its properties of any material I know, considering its vast use. On the other hand, if we consider some of the rarer non-metallic materials used in defence and aerospace, we find they have an enormous amount of development going into them. Somewhere in the middle lie iron and steel which have been investigated in some depth but rather lightly, I would suggest, in view of their total importance and their total usage.

My third point, touched on by Mr Nijkerk, is that we ought to be considering a little more carefully how we use our materials in items of heavy engineering and heavy construction. This brings us to the fundamental matter of reconsidering the materials input into engineering design codes. It seems to me that most of our heavy engineering structures, whether metallic or non-metallic, are grossly over-designed. I would not like to put a factor on this but it is probably quite significant and it seems that, as materials people, we are perhaps failing in one of our more obvious duties in not feeding enough materials knowledge to our engineering colleagues to avoid gross waste of precious natural resources.

Mr NIJKERK:
To answer your first questions about uranium and concrete: I feel like the father walking with his wife and son. When the son asked 'Why is grass green?' he replied 'I don't know'. 'Why is the underground driven on rails?' . . . 'I don't know'. 'Why is the sun shining?' Again the father said 'I don't know'. Then his mother said 'Stop asking all these questions of your father' — who responded 'Let him ask; that makes him wise'. I am sorry to say I cannot answer your problems about uranium and concrete although I can try to answer your question about heavy construction. I fully agree with you that when we construct something — a bridge or an engine or a car — we should take into consideration its future recyclability. We should not make it so complex that it cannot be recycled and, of course, we should not use too much material. Now, voices have been raised in various countries urging that Governments should award a mark or a sign for recyclability and maybe impose a penalty on products which are made unrecyclable (as I showed with my slides) like tyres. Tyres form an enormous problem but I fully agree with you that our industry and yours should take this into consideration when we make our products.

THE CHAIRMAN:
If I may comment from the point of view of the Working Party, we *did* consider uranium, but thought it rather a difficult area to get into, particularly because we had heard all these wonderful stories that if we could bring ourselves to become involved with types of

177

reactor which did not require natural uranium, if we could go on to fast breeders, there is no uranium problem at all.

Dr BARFORD: A simple recycling problem.

THE CHAIRMAN: Yes.

Mr NIJKERK: Uranium is not a commodity which is commonly being recycled, as far as I know.

Dr DYKER:
Has there ever been, to your knowledge, any recycling of military hardware?

Mr NIJKERK:
Yes, an enormous amount of recycling, especially in the United States, where the government sells tens of thousands of tons of used equipment, although there is a lot of military equipment lost during wars. For instance, in Vietnam there were about 60 000 tons of brass shells lost, which many people tried to find but were unable to do so because of landmines; war, of course, is a great spoiler of metal, notwithstanding that future wars with atom bombs will maybe use much less material.

THE CHAIRMAN:
There is a great recycling of aircraft engines, particularly military aircraft engines, and some wonderful alloys result from them.

Mr NIJKERK:
This is a very important part of recycling, as is also the aluminium recovered from obsolete aircraft.

Dr M. E. HENSTOCK
(University of Nottingham):
I can add something to that last question. I believe there are French farmers still making a fairly good thing out of taking out World War I shell cases of brass and selling them, and I also know that the United States does export non-strategic military electronic hardware to areas of low labour cost such as the Philippines and Hong Kong where they are dismantled. The components go to other parts of the world where they are built into low-cost electronic goods. If we buy certain of these there is at least a chance that some of their components started life in a B52 bomber.
The matter I want to raise with Mr Nijkerk is that some of us are increasingly worried about residual levels in materials, and since

this is a gathering of metallurgists I feel free to expand upon it, whereas I cannot when talking to economists. When we talk about steel, my specific question would be: at what point are we going to have to become concerned about widespread contamination? We are making goods cheaper by mass production but this is manifesting itself in increased residual levels. Copper and tin are particularly worrying, for they rose sharply in 1917 through increased use of scrap and they have never reverted to pre-1917 levels. Any steelmaker will tell you that copper is the element that causes him most problems on a day-to-day basis because of the effect it has in causing hot shortness in steel.

Mr NIJKERK:
You are right. Many materials disappear in steelmaking because the steel contains more and more of what in the steelworks we call impurities like copper, chromium, etc., which mostly cannot be recovered. I think it is a cause for concern, notwithstanding that in general the steel industry is the best recycler because the rejects of the industry are mostly fully recycled. Taking the cases of aluminium and copper, when industry makes rejects of these they gather them and put them into bins and they are recycled, whereas the housewife throws away aluminium wrapping, aluminium soup bags, aluminium cans, and only 20% of it is recycled. In fact it is a matter of growing concern and it needs more and more attention; thanks to conferences like this people are made aware that when they produce or use and consume, they should always consider the future recycling and the possibility of recycling.

T. J. Tarring
Markets

Sequence corrected: at the conference Mr Tarring's presentation and the following discussion were postponed to the end of Session B, under the chairmanship of Mr Dowding.

Dr R. G. BAKER
(The Glacier Metal Co. Ltd):
Mr Tarring has explained that the working of the Exchange in forming a price and providing a forum for speculation tends to damp short-term world price fluctuations. The longer-term cyclical fluctuations in prices are also serious and damaging in many ways. The Chairman has pointed out that, at present, the price of nickel precludes investment in mining. Presumably, if the demand for nickel goes up there will be a sharp price increase which will be maintained during the inevitable period of shortage covering the lead time for expanding mining capacity. When the extra capacity becomes available there will almost certainly be an over-reaction in supply and price to complete the cycle. My question therefore is: has the Exchange any possible role to play in attenuating these longer-term fluctuations? They are damaging because everyone finds it difficult to look ahead and forecast prices. Do you feel that counter-cyclical stockpiling by Government would have any effect in damping these cyclical fluctuations?

Mr TARRING:
I would disclaim any credit for having made any case for the Exchange. I am not an apologist for the Exchange, but a completely objective observer of it as a fact of life in the industry as a whole. You have referred to cyclical fluctuations by which I take it you mean year-on-year, rather than day-to-day, fluctuations. All I can say is that if you try to stop these, you will emulate King Canute and get your feet wet. These fluctuations are founded in human nature, for people are fundamentally inconstant; it is the consumer in the street who starts booms and slumps and the metal prices merely record faithfully the anticipated progress of that boom or slump.

Dr BAKER:
I do not feel that it is either necessary or even desirable to stop fluctuations, but I believe we should seek to damp them. At present the difficulties of forecasting metal costs in industry are comparable to those of gauging the speed of your car with a completely undamped speedometer!

Mr TARRING:
You have given me a lead into a splendid illustration which I enjoy using. It is most irritating when you are driving along the road to have to slow down and then speed up because of traffic conditions; your speedometer tells you that is exactly what you are doing, but if you were to try to refrain from slowing down when your speedometer showed you *were* slowing down, you would run into the car ahead of you. Conversely -- and this is where I return to the point I was making about the industry's use of these prices — if you are going on a rally, you use a Speedpilot which gives your elapsed average speed and time from the moment of starting out, and that does not fluctuate nearly as much as your speedometer. What I am suggesting in this paper is that the industry manufactures its own Speedpilot; it retains the speedometer, which is very valuable for instantaneous readings, but it applies the readings of the Speedpilot (which are derived from the speedometer and therefore cannot be false) to its direct producer/consumer physical trading activity.

Professor J. NUTTING
(University of Leeds):
I enter this argument with diffidence, but there is the point of view that the speculator always gets it wrong in that he buys on a rising market and sells on a falling market, while you could, if you were brave, perform these actions the other way round. If the speculator buys on a rising market, he forces the price higher; if he sells on a falling market,

he sends the price down. This in effect makes for volatility. One occasionally wonders whether, if we had another breed of speculator who bought on a falling market and sold on a rising market, he might make more money.

Mr TARRING:

Speculators are like bacteria: they are a necessary evil. Like the trade in its use of the Exchange, they are following basic instincts in what they do, and you will not in fact change their nature. All you could demand would be that the Exchange should so construct its contracts (and the construction of the contracts on the Exchange is of vital importance to the way it operates) that the speculators had less access to the market so that their influence was reduced; but that would be exactly the same as taking a certain drug, killing your internal flora and getting tummy trouble instead. The balance is in fact elegant because it is so natural. The speculators have exactly as much access to the market as is needed to balance the transactions which the trade, through its hedging activities, puts upon the market. If you take the speculators away, all you will get is the entire trade hedging one way in a falling market and you will be no better off.

Discussion
SESSION B: The metals

Chairman: M. F. Dowding CBE

THE CHAIRMAN:

This Session B of the conference concerns the specific metals. Instead of calling this conference 'Future metal strategy', we might well have called it 'Whither metals?' and I am hoping that, apart from hearing each author speak on his own specialty, its problems and its future, and probably its new technology, we shall hear a 1980–1985 and 1990 forecast of each metal's availability or scarcity, and particularly the author's estimate of the price at those dates.

S. J. Ross-Macdonald
Aluminium

Dr A.F. BONFIGLIOLI
(University of Sussex)

I would ask you to comment on the prospects of technical change in the field of energy saving, in particular in smelting, making a distinction between major technical changes like the Alcoa processes and technical changes in the present technology such as titanium diboride cathodes. Could you give your views on the possibilities?

Mr ROSS-MACDONALD:

I would like to put it in the perspective of the energy savings now being made. Generally, smelting of the old type using non-prebake electrodes ran at something like 17 kW/kg of metal produced. Increasingly, with the use of the prebake anode and with some better cell designs and larger cells, this has been brought down to about 15 kW/kg. Immediately before us is the process being developed by the Aluminum Company of America (Alcoa) which is looking for a reduction down to 12 kW/kg and has the potential, it is believed, to lower this to 10 kW/kg (although the capital cost of attaining those last 2 kWh may be such that it will be a case of the law of diminishing returns). Those are the technologies that confront us today. We are talking of a potential saving in today's terms of about one-third, which I think is very significant. The capital investment required to achieve this has to cover the rebuilding of nearly all the present world capacity, so that the savings will be very slow to creep into the system.

As for the totally new technologies — the sub-halide distillation methods for example — I do not think it is worth conjecturing about them because nobody has yet made a plant work effectively. Alcan came nearest to success at their Duchayne plant, where, however, they met such severe stress corrosion problems trying to keep the distillate under control that they finally had to abandon it. Both Kaiser and Alcoa say they do not think that the process is feasible with the materials we have available today. Therefore I think that something like a one-third saving in power consumption in new construction using the Alcoa process is probably the best prospect at the moment.

THE CHAIRMAN:
I was talking to Sir Alan Cottrell about aluminium some two years ago, when we were trying to predict what the future held, particularly taking account of the fact that energy would be scarcer and increasingly expensive. He was also concerned with nuclear power, and asked why all the new alumina plants could not be sited where a nuclear plant could be located without dispute; pointing out that bauxite is usually found in places 'only occupied by the lesser spotted warthog and such creatures' where one could build a nuclear station without the least controversy. Is that being thought of as a possibility?

Mr ROSS-MACDONALD:
Yes, I think it has been. Various companies involved in nuclear power construction and the supply of nuclear fuel went round the world promoting the idea of putting smelters on top of bauxite mines, having the intermediate alumina plant close by, of course. Generally, the problem has been the same one that arises in the Gulf. In Bahrain you see producer gas being used while cheap associated gas is being flared 17–20 miles away in Saudi Arabia. The problem is how to bring an industry of scale there, providing the housing and schools and all the other things that are needed; it is the infrastructure cost and the resistance of some governments to the scale of the intrusion which has made it additionally difficult. You also have to consider matching the scale of the nuclear plant with likely power consumption. I believe the smallest commercial nuclear plants are of the order of 2 000 MW; this would require a smelter of some 800 000 tonnes per annum, which is at least three times the size of the largest plants being designed today.

M. H. Davies
Copper

Sir RONALD PRAIN
(Selection Trust Ltd):
May I say first of all that I agree entirely with Morgan Davies's general conclusions about there being no danger of a shortage of copper for many years. I would, however, like to add a word of warning about the reserve figures which he showed. They are absolutely accurate according to the books he referred to, but the fact is that these copper reserves, expressed as copper as they were and not as ore, are a moving picture. There is no moment in time when they are the same as at the previous moment in time. I calculate that since January

this year the world copper reserves will increase by 50-100 million tons through an increase in price without a corresponding increase in cost, because the definition of copper reserves must be those which can be mined at a profit, except, of course, as I said earlier today, in some countries where there are nationalistic tendencies and they will produce copper even at a loss; whereas under a capitalist system the copper reserves of the world have to be measured by the relationship between the price and the cost. Strictly speaking, under the capitalist system, two of the Zambian copper companies two years ago could not have conformed to British accounting standards because they showed ore reserves of 80-100 million tons each which were not ore reserves because they could not have been produced at a profit. They should have shown no ore reserves if they had been subject to British accounting systems. This shows the complexity of calculating copper reserves.

It has been said that if someone could develop a method of producing copper at a cost of say half of what it is today, that would at one stroke increase the world's copper reserves to a far greater extent than all the exploration efforts in the world. On the last point, the general feeling that the world has been largely explored for copper is by no means true. There are vast areas in the Andes and other places in Africa where there has been no exploration. I make these random points on the one hand to confirm or endorse what Morgan Davies said about there being no need to worry about copper supplies, and on the other hand to warn against trying to quantify these as exactly as the reference books do (they are bound to be wrong).

Mr DAVIES:

I entirely agree with Sir Ronald. I obviously did not succeed in trying to inscribe the Lord's Prayer on the point of a pin, as it were. I should have spent a little more time dealing with the slides, although I had hoped that they were self-explanatory. Looking at the crucial one, showing the lifetime of copper reserves: naturally they are based on whatever estimates were made in 1960 and 1976, and I would like to draw attention to the facts which I think emphasise the point that Sir Ronald is making. In 1960, the reserves were

estimated at 154 million tons. By 1976, they had risen to 451 million tons. I should think that by the year 2000 they will be in excess of 500 million tons — and that is a very conservative figure. We should not, as economists tend to do, consider these numbers as Holy Writ. It is a fluid situation the whole time.

Sir RONALD PRAIN:

I quite agree, but I would not have thought that mere increase in price was going to increase copper reserves. Decrease in costs can be just as important, and I gained the impression from one of the graphs that some of the increases in reserves were related to an increase in the price of copper. This is the point I am making. The increase in reserves is a function of two things: discovery rates, which are calculated quite closely by the mining companies, multiplied by the difference between price and cost.

Mr DAVIES:

I think Sir Ronald and I are in absolute agreement. The 1960 figures are from the US Bureau of Mines. There have been figures in the intervening years, but that is the most recent reliable estimate of the world reserves and was in fact based on the UN study. It is very similar to the estimate made by Krauss earlier in 1976.

Mr A. A. ARCHER
(Institute of Geological Sciences):
I would like to reinforce strongly what Sir Ronald Prain said about the influence of both prices and costs on reserves. In that context, Mr Davies quoted a figure for reserves of copper in manganese nodules but at the moment there are no reserves in manganese nodules because the cost of producing copper from nodules is unknown. There are what can be described as potential reserves, which is rather different.

I would like to ask two questions. I was very interested in the table Mr Davies showed for per-capita consumption of copper in different countries and was slightly surprised to see that the figure shown for the United Kingdom is about the same as for Japan and the USA. As used in that table, how is consumption defined?

My second question invites Mr Davies to go just a little further and give us his opinion on what may be the impact of optical glass fibres on copper consumption, bearing in mind, as he said, the importance of electronic and communication uses of copper. Does he think that glass fibres could make a sensible impact on copper consumption in the communication field?

Mr DAVIES:

Firstly, I too would like to know the sources of some per-capita consumption figures, which in my case came from the UN study. The total numbers are reasonably reliable, although I do not think the base for total consumption (by end-use) in all the countries relative to populations are the same. There is the question of classification, for example, but I do not think it upsets the main argument. In the industrialized regions in Japan, the UK, Germany, and the USA, the per-capita consumptions are all about the same level. The figure that I showed for 1972-74 shows Germany high at 11·5 and the UK, Japan, and the USA all between 9 and 10. Those per-capita figures, however, are confused by the fact that they ignore copper exported in the form of finished goods, motor-cars, electrical appliances, etc.

Turning to fibre optics, perhaps I should make the excuse that I do not have my fibre optics crystal ball here! They certainly are going to be used. Present predictions are that main trunk exchange links will use fibre optics extensively and trunk/local exchange links very substantially. It really is a question of system costs and not simply a question of scarcity and price of materials though these are linked.

A year or so ago, the Post Office made the spectacular announcement of their decision to switch to aluminium for local distribution cables. I think that, in tactical terms only, they have since regretted that decision because of the relative movements of price of copper versus aluminium in that time. Whether they will prove to be right in the long term I do not know, for we have heard the proponent for aluminium expressing fears for its price levels in relation to its energy sensitivity as the years go by.

Professor J. NUTTING:
(University of Leeds):
I think I nearly obtained an answer to my question but if one looks for the rival (as it were) to copper, it is aluminium. Accepting the arguments that with both aluminium and copper there is no shortage of material, there is good evidence that the use of a metal is related to its price and therefore the argument is that if copper increases in price faster than aluminium, then this is going to favour substitution of aluminium. Perhaps the weakness of the organization of this meeting is that we are discussing metals individually whereas the problems with which we shall be faced are the future variations in price relativities.

Mr DAVIES:
Sir Ronald has highlighted one of the problems in talking about price. We do not at present have a system in which the price of commodities in general, let alone copper, is controlled solely by market forces. At times during the period of low prices in the last four years public statements were made that the cost of producing copper in Zambia and Zaire caused a net foreign exchange loss to those countries. That state of affairs clearly cannot continue. As Sir Ronald pointed out, mineral resources are very important to developing countries; they will continue to produce them; and they will continue to influence world market prices, but partly from national requirements not necessarily related to market economics.

THE CHAIRMAN:
I believe £1 400 a ton was mentioned by Sir Mark Turner as the figure at which people would start mining again. What are your comments?

Mr DAVIES:
It is the same question, if I may say so, as 'How long is a piece of string?' It depends really on the ore body itself. There is a substantial variation in the cost of recovery of refined metals from different ore bodies. The management of particular resources will assess what is the right time in relation to the prospective cost to bring a major mine on stream.

D. A. Robins
Tin

Mr B. G. THUNDAL
(Granges Aluminium, Finspong, Sweden):
I would like to ask about tinplate. Aluminium is a very strong competitor to tinplate, especially in canning. What do you think about the future part that aluminium will play in the canning industry compared to tinplate?

Dr ROBINS:
The role of aluminium in the canning industry is, in my view, a relatively small one. It is of considerable importance to the aluminium industry but only competes with tinplate for beer and beverage packaging. Aluminium is not used to any significant extent for canning food. One of the reasons why aluminium has captured a substantial part of the beverage can market has been the ability to make two-piece D & I cans. Now that this same process can be applied to tinplate, however, I feel that the use of aluminium is likely to decrease. It should be noted that the second largest aluminium producer in the USA is installing a canning line to make two-piece tinplate cans.

Dr R. G. BAKER
(The Glacier Metal Co. Ltd):
It seems to me that Dr Robins may have made two mutually contradictory predictions; a progressive increase in the use of tin, and a steep progressive rise in its price. The price curve reminds me of one of the more fearsome illustrations in 'Limits to growth'. How long must we anticipate that the apparent exponential growth of the price curve will continue? Dr Robins must presumably expect that it will level off in the light of his prediction that the usage will increase. I suppose the answer to my question depends on the possibility that increased efficiency in techniques of production will give the increas-

ed return to producers which they might otherwise expect to gain from a continuously accelerating price increase. If present trends continue, people are going to look for substitutes, I would have thought.

Dr ROBINS:
I did question the value of showing the curve on tin prices, since in some ways it is misleading. It is not possible in the time to discuss it in detail, but the working of the International Tin Agreement kept the tin price at too low a level for a long time. The price then broke out of the ITC range. If, therefore, we had looked at the price over a longer time-span then the overall increase would have been less. There is no evidence that such a rapid rise in price will take place in the near future. Indeed, with the approximate balance between supply and demand coupled with the threat of releases from the USA stockpile, a relatively weak price might be expected for a short time. The point I would like to make is that it will not come down significantly and in the longer term is likely to increase. This also means that in the long run it will be used more efficiently.

Dr BAKER:
In defining the reserves of 100–150 years, were you referring to 'economically exploitable reserves'?

Dr ROBINS:
The figure of 50–150 years was that given by Professor Hosking. Proved reserves tend to cover 15–30 years since mining operations in a particular country do not normally bother to explore further ahead than this. The tin price also influences reserves, since at a higher price level otherwise uneconomic ores become commercially exploitable.

THE CHAIRMAN: You put 45 years in your paper last year, did you not?

Professor J. NUTTING: Yes, but that was not a very authoritative estimate.

Professor W. O. ALEXANDER
(University of Aston in Birmingham):
There are two threads running throughout the papers given so far this afternoon which worry me. None of them have paid any attention to the total energy content and the obvious rise in price of total energy. The 'Aluminium' paper referred briefly to it as a

matter for concern, but of course this is the one great asset of aluminium, that if you can recycle a high proportion, then you have virtually a combined energy and material bank. There is one other aspect of both aluminium and tin (and several other materials, particularly plastics) which again, in the longer term, I think we must worry about; oddly enough, they are mainly related to food usage and the way we buy our food and package it, and also, to a lesser extent, to certain other developments in composites. It is a fact that, whatever the incremental energy requirements are at each stage of processing a metal or a plastic, with perhaps one or two exceptions of certain plastics, the end stages of film, fibre or filament formations are heavy in energy usage. I think this is another long-term question that wants looking at. Can we justify the formation of all this waste surface area, particularly if you are going to have great difficulties in getting anything like 100% recovery as old scrap?

Dr ROBINS:
I believe that we are in fact taking into account energy considerations when we talk about price. Some of the recent price increases for tin are undoubtedly related to the increase in the cost of energy required to handle the very lean ores. As the price of energy goes up, so does the cost of tin production, and this is reflected in the price. As the producers start to get a better return for their tin with higher real price levels then this will encourage more efficient use.

THE CHAIRMAN:
It occurs to me that so far we have heard speakers on three metals all of which to some extent look for substitution one for the other in their various roles. I was not on the Steering Committee for this conference, but it might have been better to have asked the 'copper' man to have spoken on aluminium and the 'aluminium' man on copper, then inviting Dr Robins (tin) to have spoken on both. In that way, I think, we might have heard the real future that these metals may face. Each of the three has given a most learned and accurate paper giving his own view on his own metal, and naturally — as I would do it if I were in their place — one sees the future as fairly good. Let us hope it will prove to be so for all of you.

A. R. L. Chivers
Zinc, cadmium, and lead

Dr R. G. BAKER
(The Glacier Metal Co. Ltd):
I did not hear you make any forecast of the price of lead. Would you care to do so?

Mr CHIVERS:
I do not really think I can forecast the price of lead. It is subject much more to fluctuation than the zinc price. The ratio between the price of zinc and the price of lead has varied a lot in recent months.

P. L. Dancoisne
Manganese

THE CHAIRMAN:
You said that for the total tonnage of steel in the world, which is now about 700 million, the consumption of manganese was about 5 million.

Mr DANCOISNE:
A little more than 5 million.

THE CHAIRMAN:
In an address I gave last year I put it at 7 million — being a little optimistic on your behalf.

Mr DANCOISNE:
Were you considering the manganese alloys or the manganese in metal content?

THE CHAIRMAN:
I believe my figure was for the content and perhaps I was putting it too high.

R. B. Nicholson and
P. G. Cranfield
Nickel

Dr C. D. DESFORGES
(Engelhard Industries Ltd):
To what extent are the prices you have just quoted offset by the precious metal recovery from your deposits?

Dr NICHOLSON:
The prices are in no way related to the recovery of precious metals from our deposits; the costs are related, but I did not quote any costs. It is probably known that only the sulphide deposits have precious metals as a by-product, and so the vast majority of the world's reserves do not have precious metals associated with them. Typically, I would say, the sales value of precious metals as a percentage of the sales value of nickel has been of the order of a few percent, being substantially higher, of course, for copper and cobalt by-products.

Mr A. A. ARCHER
(Institute of Geological Sciences):
Dr Nicholson, like previous speakers with respect to their metals, has given us a rather comforting picture of the world reserves position and the capability for meeting demands in the future. I wonder if, however, it is not perhaps slightly too rosy. If I noted correctly, he said that world consumption is about half-a-million tons and world reserves about 75 million tons, ending with a simple calculation of about 140 years' life. My impression is that the half-million tons consumption refers to the Western world and it is also my impression that the figure of 75 million tons is world reserves including the centrally controlled economy countries. Furthermore, no account is taken of the 4–5% growth rate which he predicts, which results in a doubling about every 16 years; so I would

have thought that 140 years is much too optimistic in the light of our current knowledge of reserves.

Dr NICHOLSON:
You are absolutely right on the first point. In my anxiety to try to make up some time for the President, I did not put in any of the caveats I should have done. The consumption figures refer to the non-communist world, whereas reserves referred to total world and included a significant quantity in Cuba and USSR. Although the consumption in the communist world is still quite low — I think we would probably guess that it might be somewhere around 20% (not more) of that of the non-communist world, it reduces 105 years down to something like 85. On the growth rate question my comments still stand. That is why I quoted the figures for 1950. The point is that responsible mining companies develop their reserves at a rate which strikes a happy medium between spending shareholders' money unnecessarily and at least assuring their customers that there is enough metal available. In 1950 it was felt necessary to prove out 14 million tons of reserves; in 1978, 75 million tons of reserves. If by the year 2000 nickel is being consumed at the rate of 1 million tons a year, I have no doubt the reserves will be around 150 million tons — it is as simple as that. There is then the additional point about deep-sea nodules, which have a significant nickel content as you know. I think, however, that even without deep-sea nodules one can still face the foreseeable future with complete confidence as far as nickel reserves are concerned.

Mr ARCHER:
A quick riposte: is it not the case that the 75 million tons of reserves are not proved but inferred reserves?

Dr NICHOLSON:
These are US Bureau of Mines data and the 75 million tons are described as proven and probable.

THE CHAIRMAN:
Mr Archer was not taking account of the fact that the original major nickel discovery in Canada was a piece of happenstance arising from building a railway.

Professor J. NUTTING
(University of Leeds):
It is perhaps not surprising that those whose
livelihood is built around one particular metal
should publicly adopt an optimistic stance
when discussing that metal; but I wonder how
Dr Nicholson views the possible substitution
of nickel by manganese in, for example,
austenitic steels?

Dr NICHOLSON:
It is perfectly true, as I have pointed out, that
nickel consumption depends very heavily on
austenitic stainless steels and there is a
possibility of substitution both by manganese
and, of course, by nickel-free ferritic steels.

It is a fact that the proportion of nickel-
containing stainless steels has remained at
about 70% for the last 30 years. I cannot
explain why it has done so, but there seems
no real reason why that proportion should
suddenly change unless the nickel price went
out of sight compared with other metals (and
I see no reason to expect it to do that).

A second point about substitution: the
nickel industry has lived with substitution
since 1920, when the bottom dropped out of
the market at the end of World War I; the
major new application was then nickel-
copper alloys (the proportion of nickel used
in Ni–Cu alloys is nowadays rather small, yet
that was the major application). So historically the producing companies, the users, and
the research institutes have been able to develop new applications as each old one faded
out for whatever reason, and we expect to
continue to do that.

Sir RONALD PRAIN
(Selection Trust Ltd):
Several speakers have referred in their papers
to reserves and I do not apologize for raising
this question again because, in a way, it is the
theme of this whole meeting — the future. It
is significant that one or two speakers have
shown how reserves have increased as time has
gone on where they might have been expected
to decrease.

I would like to make three points on this.
In my opinion, it is a function of time rather
than price because, secondly, no mine proves
up more ore than it has to in order to justify
the initial installation of a service plant. The
third fact, which is probably far less known

here, is the depletion laws of the United
States, which for many years made it a
positive disadvantage to show one ton more
ore than absolutely necessary. The whole
taxation system was based on the life of a
mine, and therefore the shorter the life the
better, which is why mines in the States and
in other countries do not look for ore until
they have to. I think it is very important to
bear in mind when discussing the future that
there are endless millions of tons of ores of
various metals all round the world in the
existing mines of the world which are not
going to be shown until they have to be.

THE CHAIRMAN:
It is delightful to be able to blame the
Treasury yet again. You have made a very
strong point.

Professor ALEXANDER:
I wonder if your bet is still on, Dr Nicholson,
if you allow for the labour and energy
content of all your capital expenditure.

Dr W. E. DUCKWORTH
(Fulmer Research Institute, Stoke Poges):
It appears from the discussions so far that
there will be no significant metals shortages.
Energy scarcity does not appear to worry us
either, because that will be reflected in the
metals price. However, if we are going to open
up all the new reserves of metals that we have
been hearing about a large amount of capital
investment will be required.

In our increasingly consumer-oriented
society the ratio of investment money to consumption money has been falling and I
wonder if we may find ourselves not having
enough investment money available to exploit
these reserves.

Dr NICHOLSON:
I think that is very possible. There is obviously
competition for the available investment
money and if the return from a metal mining
project is sufficient, then that project will
win the competition. The point I was making
was that, at the moment, the price of nickel
(and obviously of some other metals— gives
an inadequate return and therefore you will

get a shortage due solely to starvation of capital.

Coming back to Professor Alexander's point, I think that one reason for the very high capital cost of new mining projects is that they tend to be in remote areas, and therefore a significant fraction of the cost is for providing infrastructure. You are making homes for the population who did not have them previously and I am not sure how that is entered into Professor Alexander's energy balance; you are also making roads and putting up hospitals.

Professor ALEXANDER
(written reply):
One must of course include the total energy and labour costs of the infrastructure.

Mr N. WHITTER
(BNF Metals Technology Centre, Wantage):
Dr Nicholson suggested a possible increase in the use of nickel in batteries. I wonder whether he was thinking of the possible use of nickel-containing batteries for vehicle propulsion. This is important in the light of Mr Chivers's remarks about the future of the lead–acid battery.

Dr NICHOLSON:
I based my comments partly on history, which shows that the use of nickel in nickel-cadmium batteries has been one of the most dynamic growth areas in nickel, although because it started from a very small amount it is still relatively modest. One of the probable reasons is the big increase in rechargeable batteries for various consumer applications, accompanied by a similar increase in batteries for stand-by applications; the Health and Safety regulations have stimulated demand for stand-by batteries. Thus every cloud has its silver lining; but, clearly, the major market which could become available would be the vehicular battery, for after lead–acid, nickel–iron and nickel–zinc are regarded as possibly the next generation of batteries.

A. M. Sage
Vanadium

THE CHAIRMAN:
We have to thank Dr Sage very much because, as he told us, at very short notice of only a week he and his colleagues have produced an extremely clear and detailed picture of the vanadium situation in the world. When historians come to look at our records, I think it will be the first forecast in which we actually mention the years beyond 3000 A.D.

Dr D. S. FLETT
(Warren Spring Laboratory):
A comment, first of all. It is very interesting to note that in Sweden recently there has been a new process developed for recovery of both nickel and vanadium from fly ash and from soot. This process is not surprisingly called the 'Sotex' process. What is equally interesting, I think, is the joint venture of the company involved here, MX Processor, in collaboration with a shipbuilding company to produce a custom-built barge on which a plant has been erected. They aim to sail up and down the coast of Sweden processing soot and fly ash. There is a land-based plant already in operation but I have no information as to how it is actually performing. In this respect, it is proper, of course, that you mention the Athabasca tar sands as an unworked resource. I could not read your figures on that. How much vanadium is available there and what percentage of the unworked resource is it?

Dr SAGE:
150 ppm, and the reserves are estimated at 14 000 million pounds V_2O_5.

P. J. Kiddle **Tungsten**	S. R. Keown **Niobium**

THE CHAIRMAN:

What you said about the Chinese situation was very interesting. It underlined to us (and there are two or three members of our Chinese Mission here) that tungsten was one of the things with which they expected to be in surplus and pay some of the bills for the enormous imports we hope they are going to make in other directions.

Mr M. H. DAVIES

(Copper Development Association):

Since there are no questions, perhaps I could be permitted to make one or two points. It seems to me that, during the whole of today, some of the questions at least appear to be based on a misconception, perhaps because the underlying theme of the conference itself is based on a misconception, i.e. the reputed shortage of supply of metals. I believe that the facts presented by almost every speaker today demonstrate that there is no actual shortage and no foreseeable shortage. May I also anticipate some of the remarks that might be made tomorrow in reiteration of some made today, by suggesting that there is no need for The Metals Society to contemplate a change of name to become the 'Timber and Concrete Society'!

THE CHAIRMAN:

We take note of what you say. I agree that there seems to be a certain atmosphere of complacency and smugness beginning to creep into the present conference. Nevertheless, I think the architects of this conference still believe that out of the whole discussion will come some realistic views on what may need to be done to justify this complacency. Tomorrow there will be an opportunity to do some summing up, as Mr Davies suggests, and we look forward to that in the latter part of tomorrow morning.

THE CHAIRMAN:

It is always good to hear a real enthusiast talking on his subject. It is also good to hear about the metals which are going to serve the microalloys which are going to be so important, particularly those that I think (as a veteran in rolling) do not require the rolling process to be interfered with as happens when we have to do temperature-controlled rolling, for example.

Dr R. B. NICHOLSON

(Inco Europe Ltd):

May I ask what the metal cost of a 10 pence piece made of niobium would be?

Dr KEOWN:

A 10 p piece weighs something like 12 grammes and the metal cost is at the moment roughly four times that of cupro-nickel (which I think is 75% copper and 25% nickel); so we have a 4:1 cost problem at present, although it seems sensible to try to replace copper and nickel and use nickel for other things. Also, of course, coins do not need to have the same weight as in the past; our 10 p piece is clearly a fairly heavy coin. One can look at it in various ways. If one looks at the coinage possibility, why should we use certain elements for coins when they can be used for other things? The cost in fact would be between 5 and 6 p for a 10 p coin, as against about 1½ p at the present time. If you look at the 50 p piece it becomes more reasonable, because a 50 p piece does not weigh five times as much as a 10 p piece (I use 10 p as an example because we are all aware of it). We have one country in the world which is considering introducing niobium as a currency material. Some countries have different materials in their coins and different weights of coins. However, I do not think anyone would consider introducing niobium without a very good economic reason.

A DELEGATE:

There will very shortly be, I am sure, a demand for a £1 coin. I think this is a possibility which might interest people.

J. Nutting
Cobalt and chromium

Mr J. C. LEACH
(BSC Stainless):

I was pleased that Professor Nutting referred to the difficulties which may arise in guaranteeing continuity of supplies of chromium, and his quote from the recent West German analysis of how a significant reduction in chromium availability might affect that nation's economy brought home particularly how serious an industrial problem the Western world would face were Southern Africa's political problems to affect raw material supply. As a member of BSC Stainless, I represent by far the largest consumer of chromium in this country. Some may remember that a few years ago permission was sought from the government here to invest certain accumulated capital held by BSC in South Africa in a new plant then being proposed for production there of ferrochrome; and how this request was turned down several times before finally being authorized by a new minister. That plant is now working very successfully and not too long ago we took the first bulk delivery from it into this country and into our works.

We have naturally been aware for some time of the dangers that exist through being so dependent for a major and indispensible raw material in stainless steel manufacture upon a single source. It was interesting to hear Sir Ronald Prain's remark to the effect that when demand increases, new resources tend to appear. We realise that other steelmaking nations have surveyed in some detail the map for new chromium reserves but without signal success. True, deposits more sizeable than at first thought were found in Finland, but generally in the world at large no new major deposits have come to light. So we are still left with the preponderance of raw chrome in Southern Africa, some 25% of it in Rhodesia, and over 70% in South Africa, and it may be timely to remark that although BSC has certainly taken considerable quantities of ferrochrome from South Africa it has, unlike all its major overseas competitors, not knowingly taken since UDI any of its chrome from Rhodesia.

Mention was made in earlier papers of the impact which the new stainless steelmaking processes have had upon the consumption of the various types of ferrochrome used. AOD is now far and away the major process used in the West for the production of stainless steel. This process has at a stroke increased the steelmaking chromium yield from raw material to liquid finished steel by at least 10% (say from 86% to over 96%). Moreover, steelmakers are now able to use without difficulty and without penalty the cheapest form of alloy chromium units available, namely those with high carbon, and sometimes high silicon, contents; and these are relatively cheaper because in their manufacture less than half the energy is required per unit of chromium that was necessary in the manufacture of the low-carbon alloys, so vital in pre-AOD stainless steelmaking.

This leads to another point, which has perhaps not been emphasized sufficiently at this conference: that the conservation of alloys can be helped by maximizing their through-yield from ore to liquid steel. In other words, improvements both in the manufacture of ferrochrome, for example, and in its use in steelmaking should be examined together to see if there are further means of increasing this through-yield. At one time the chromium yield from ore to ferrochrome was no more than 80% and this with the steelmaker's chromium yield of 86% gave a through-yield

191

of less than 70%. Now, a yield of up to 90% in ferrochrome manufacture may be attainable, which with the 96% yield of steel-making gives a through-yield of over 85% — high, but still leaving room for improvement by joint efforts.

Mr A. A. ARCHER
(Institute of Geological Sciences):
I would like to reinforce what the previous speaker has said. Since the days of Agricola, the mining industry has not been inactive in exploration, and what may have applied in the 14th century may not apply now. The speakers who have talked to us so interestingly about copper, cobalt, nickel, and manganese have mentioned manganese nodules, with which I have been rather preoccupied for the last 10 years at the United Nations. I would not like your meeting, Mr Chairman, to get the impression that access to nodules is going to be as free as some people would like to think. It is agreed by all nations in the Conference, which is still continuing, that the resources of the deep-sea bed will be controlled by an International Seabed Resource Authority, that some of the mining will be done by a United Nations mining company called the Enterprise, and that access by companies, the companies who are developing the technology now, will be regulated. In particular, at the insistence of the land-based producers of those four metals, at present the text contains a production limitation formula which will maintain production of these four metals from the sea bed below the level that might be sustained by allowing free market forces to apply.

If nodule production gets going as so many of us hope it will, this could lead to an over-supply of cobalt in terms of present consumption. The cobalt-to-nickel ratio in nodules is about 1:5. This will mean a great increase in availability of cobalt and most projections lead to the conclusion that this could lead to the price of cobalt coming down to about the same price as nickel. I would like to ask Professor Nutting what metallurgists will do, if they find cobalt available on the market at the same price as nickel.

My second question pursues the same line as a comment Sir Ronald Prain made yesterday. A slide shown by Professor Nutting of static resource life for different elements gave a figure placing cobalt in the same category as iron and manganese with a life of more than 10^4. Could I ask him what definition of resource he is using?

Professor NUTTING:
Firstly, with regard to manganese nodules, let us not be politically naive. My guess is that manganese nodules will be exploited. The nations that have supplies of manganese at present may well be able to exert some control at the moment in avoiding flooding the world markets with manganese. But once the advantages of exploiting nodules, not only for the manganese content but for the other metals they contain, is appreciated then it is doubtful if present agreements will be honoured. The political leverage of the present manganese agreements is not all that great and technical developments are likely to lead to reappraisal of international agreements.

If, as a result of exploiting sea-bed nodules, there were to be a glut of cobalt, then there are many uses for the metal. In many applications it could replace nickel. But cobalt has an allotropic change and, in this respect, it is metallurgically more interesting than nickel. It is possible that high-strength alloys could be developed on a cobalt base, by making use of this allotropic change. However, it has been shown that if the real price of metal is halved, then the consumption may go up by a factor of three. In the past the cobalt/nickel price ratio has been about 2:1. The present price ratio is distorted because of political factors, but if the price ratio did fall to 1:1, then I believe that there would be considerable scope for the development of new cobalt-containing alloys.

Finally, the estimates of land-based cobalt reserves are very much open to doubt. In relation to the crustal abundance, the estimates seem abnormally low, and this could be an indication that there are still extensive cobalt deposits awaiting discovery.

Sir RONALD PRAIN
(Selection Trust Ltd):
Since this meeting is concerned with future strategy and you cannot eliminate stockpiling from considerations of strategy, perhaps we could spend a few minutes on some aspects of

stockpiling. I personally think Professor Nutting was right to suggest that the German idea of stockpiling chromium may not be the answer. You may know that in America there has been very serious consideration given to stockpiling for the minerals which come from South Africa (chrome, manganese, platinum and vanadium) and this idea has been found impracticable so far, for two reasons. One is political and the other economic.

Politically it is felt that if America now bought for stockpile, which by definition means buying in excess of current requirements, it would create an enormous boom in South Africa which is politically unacceptable to certain sections of the American Congress.

On the economic side, it has to be questioned whether the South Africans are going to increase their capacity just to oblige the Americans for stockpiling, only to be burdened with excess capacity when the stockpiling operation is finished.

For these two reasons there is at the moment (unless something has happened in the last month) no inclination to stockpile by excess buying from South Africa. The conclusion has been reached that the only time you can stockpile, whether it is from South Africa or anywhere else, is during a period of really excess production capacity; if any country had wanted to stockpile copper, they had the most marvellous opportunity during the last three years — it is now too late. Stockpiling is the problem. That is why I venture to say that I think Jack Nutting is right to doubt whether the German policy is capable of being carried out, especially if it is going to be initiated by others. You may have seen in this week's *Economist* a letter addressed to the new Government suggesting that the first consideration they should give is to stockpiling metals for this country. This may be possible in some metals where there is an excess productive capacity somewhere in the world but I do not think it can apply to the South African minerals in the present state of the market.

THE CHAIRMAN:
I think we all agree that yours is a very important contribution to the whole sense of the meeting which we will need to bear in mind in our summing up.

Dr C. D. DESFORGES
(Engelhard Industries Ltd):
Professor Nutting mentioned two solutions: one geological — 'Let's look for more of the stuff', the other political — 'Let's be friendly with the people who have got it'. Would he care to comment on the possibility (since we are concerned with overall strategy) of the development of a coating technology which will minimize the amounts of these scarce materials of high value; furthermore, as a longer-term issue for high temperature, would he comment on the likely availability of ceramic materials which could withstand the high-temperature processes he has been describing?

Professor NUTTING:
Coatings are one obvious way of conserving materials of limited availability, but as Professor Alexander pointed out earlier, coating technology is frequently energy-intensive and this may offset the materials saving advantages.

My own view is that coatings could be more widely applied. In many cases we tend to lose sight of the advantages of coatings by putting too stringent a requirement on the properties of the substrate. An example of this is the chromizing of steel, where one of the major applications could be in the production of motorcar silencers, yet the process is not generally used because of the existing specification for the strength levels of the unprotected steels at present in use.

In relation to chromium use in high-temperature materials, Dr Desforges raises the interesting point of ceramic material substitution. In some respects we have been 'spoilt' through having metallic high-temperature materials which were strong at high temperatures and were still ductile at ambient temperatures. Our design philosophies have taken this point for granted. If we are to use ceramic materials, then we have to accept that these are likely to be brittle at ambient temperatures and therefore we have to modify our starting and stopping, i.e. our heating and cooling procedures. For aircraft engines design modifications to meet these requirements may be difficult but for land-based turbines they could be more readily overcome. Ceramic blade substitution might be possible in such cases.

M. T. Herbert,
G. J. K. Acres, and J. E. Hughes
Platinum, gold, silver, and mercury

Dr W. E. HOARE *(Consultant)*:
Could I ask Dr Acres for his opinion on the possible effect of the use of videotape recording in place of cinematographic film on the demand for silver by the photographic industry?

Dr ACRES:
It would appear that with the advent of photocopying techniques, videotape records and other new technologies in the last few years, one might have predicted that they would have a significant effect in the photographic industry on the demand for silver; but the figures do not bear this out. When one looks in detail at the various uses of silver in the photographic industry, one finds that the major uses for silver occur in industrial and medical applications, where newer technology has not yet made much impact. In fact, the definition and other properties that one expects from silver have not yet been achieved by any other means. I think, therefore, that what is happening is that the demand for silver in the more traditional sectors of the photographic industry is increasing, while the new technologies are not competing.

Dr M. E. HENSTOCK
(University of Nottingham):
Ten years ago, if one wanted to make a calculation one had to get help to carry a machine weighing about 20 kg, most of which was recyclable metal, down the corridor. These things have got smaller; I have just been given a calculator the size of a credit card and very little thicker. Clearly, it does not contain very much worthwhile metal; but these devices are now so cheap that everyone has them and it will soon be cheaper to throw them away (if indeed it is not already so) than have them repaired. On a technical level, is there any way in which one may recover the valuable materials from such a complex object, and is Johnson Matthey looking into this possibility? I have heard it said that each of these contains 35 cents worth of precious metals. Can we afford to throw away millions of 35 cents every year?

Dr ACRES:
That is a most interesting question, and when you ask whether we are looking into it, the answer is 'Yes, across the board'. The point about the size illustrates why the demand for gold and also silver in the electronics industry has apparently not increased as the number of these devices on the market has increased; the reason, of course, being that advances in technology have considerably reduced the amount of precious metal used in a single device, so that although the number of devices has increased, the amount of metal in each one has been reduced substantially, leaving demand at more or less the same level.

There is a major problem in that situation, one which applies equally to the use of platinum group metals in the car industry: that while the amount of metal per unit is very small, the total amount consumed in the world is very large and the amount of metal that can be generated as scrap, certainly in the electronics industry, can be substantial. In the case of the scrap (which in America includes the precious metals that somehow or other get into sewage waste) it is certainly recoverable, and the facilities exist to do so if the economics are right. In the case of individual units, more particularly the car catalyst units, it depends, as it does I suppose for the glass industry collecting bottles, on whether someone is prepared to organize the collection of individual units so that sufficient material is available in one or two centres to make it economically viable to recover the metal. In the case of the car industry, discussions continue as to ways and means by which, say, the precious metal in the car catalyst converter can be recovered, just as lead is recovered from batteries. I would not like to claim at the moment that there is 100% success in recovery of the metal used in these applications on a global scale but some aspects of recovery are in operation and discussions are taking place as to how to achieve greater success.

Professor J. NUTTING
(University of Leeds):
Mercury is one of the few metals that have
been mentioned in which the resource life
may well be within the lifespan of some of us.
If one then looks at the uses of mercury, a
critical one, I think, is in dental amalgams. I
wonder if you see any prospect of a substitution in amalgams?

Dr ACRES:
That, in fact, is an important area in two
respects: not only the question of availability
but also environmental and health considerations concerning people who use mercury-
containing amalgams in the dental profession.
As I understand it, gallium is being considered
as an alternative to mercury in this industry
and there may be situations where such an
alternative will be forced upon the industry
for reasons of health. In some areas this has
already happened.

Mr J. C. LEACH
(BSC Stainless):
May I ask Dr Acres to clarify a figure in his
preprinted text? He says: 'Total production
in 1977 was 240 000 flasks (110 tonnes)'.
That should be a lot more, should it not? He
says that each flask contains 76 lb. Is 110
tonnes the yearly production of mercury?

Dr ACRES:
If our arithmetic is right, yes. I believe the
figures are right. I will check them with you
afterwards.

Mr L. F. DENARO
(Consultant, Nickel Industry, London):
Ruthenium: there must be very substantial
stocks in producers' hands, accumulated over
the years. What new, promising applications
are there?

Dr ACRES:
You may be aware that in recent years, again
for energy considerations, ruthenium has now
found a major application in the manufacture
of chlorine and caustic soda, in other words
the chloralkali process. Ruthenium-coated
anodes are now widely used in at least one of
the processes used in the production of
chlorine and may make a major contribution
to energy conservation. It is possible in fact
that in future the other electrode in the chlorine cell will become platinum-coated, again

for energy conservation reasons. In addition
to that role, now well established, there are
those who believe that ruthenium has a part
to play in the production of synthetic fuels,
synthetic natural gas and liquid fuels from
coal, and so there may be a major application
for ruthenium on the horizon in addition to
chloralkali.

T. W. Farthing
and R. E. Goosey
Titanium

Mr M. H. DAVIES
(Copper Development Association):
I worked on titanium in the United States in
the early fifties, an early love which I am
afraid has been completely renounced for
copper. Mr. Goosey referred quite rightly to
the excellent corrosion resistance of titanium
in seawater, but corrosion resistance is not the
only important requirement. He refers to its
application on North Sea oil rigs where in
recent times the fouling problem, because of
the increase of loading on the rigs and plat-
forms, has become quite significant. A similar
problem arises with respect to the fouling of
condenser tubes. My question is a direct one:
what can Mr Goosey say about the relative
antifouling resistance of titanium compared
with copper–nickel alloys?

Mr GOOSEY:
To the best of my knowledge, and I must
confess this is not my particular area, there
are no significant problems with fouling of
titanium but I would hesitate to compare it
with the copper-base materials.

Discussion
SESSION C: Influence of design factors on future utilization of metals

Chairman: Dr R. B. Nicholson

R. L. P. Berry
and L. Whalley
Recycling of metals

Mr R. G. PURNELL
(Delta Metal Co. Ltd):
On the last page of the paper as preprinted, there is a reference to the role of government and the fact that the government works with local authorities to encourage the collection of domestic scrap. Has anything been done about educating children in this respect? — the identifying of cans from tins and coloured bottles and so on. I cannot imagine that the average householder is going to be very keen on putting little bits in one bag and something else in another, nor the average dustman on collecting six bags from each household.

Dr WHALLEY:
I agree entirely about the problem of inducing householders to sort different types of waste into separate bins and get them collected. There are many difficulties associated with

that idea. Certainly something has been done about trying to educate children in this regard. There was an experiment carried out last year with the help of the National Anti-waste Programme and the Aluminium Federation, in which schools in Buckinghamshire were encouraged to collect aluminium over a period of a few months; the various collections were brought together and sent to the aluminium melters who melted it down and analysed it and found that it was a very good grade of scrap that they were getting back. But the quantity was small — only a few tons over a few months. Nevertheless it was regarded as an encouraging venture, as there is a lot of energy and enthusiasm available among the population, among schoolchildren, among people who are trying to collect money for charity. I have come across all sorts of people who run these voluntary collection schemes, and certainly the National Anti-waste Programme is doing its best to encourage this.

Professor W. O. ALEXANDER
(University of Aston in Birmingham):
A propos of what I was referring to in my earlier lecture, I would have thought that Warren Spring — since they did work rather similar to mine — and certainly the anti-waste organizations would at least make a real

attempt to separate old scrap from new scrap. Those very encouraging figures that were shown for lead and steel could equally well be used to say that there is gross inefficiency, i.e. low yield, in the lead and the steel industries, because you did not know how much of the 60% load was new scrap and how much was old scrap. From the point of view of human beings attempting to recover metal that has been in service, we must first of all have real, meaningful figures on old scrap. I am sorry to have to re-emphasize it but the plain fact is that in this country there is something like 14 million tonnes of copper lying around somewhere yet no one knows where it is. Again, some of those figures conceal the fact that they include industrial residues which do not appear in the figures for the overall consumption of metal in any one year, giving an augmented apparent figure of resources which makes the whole situation worse. My plea is: from now on let us try and separate old scrap, which is all we are concerned with — the metal that has been in service. We are not concerned with new scrap.

The other problem with new scrap is (as was touched on with stainless steel) that if you recycle it within a works or with a customer rejecting new process scrap work, you may use much more energy continually recycling that rather than making a special effort to get very high yields for the first cycle only.

Dr WHALLEY:
May I comment on that? I think we know where quite a lot of this old scrap is. In the case of copper, for example (and all the other non-ferrous metals), a considerable quantity of this material is going to the scrap industry, being collected together with ferrous scrap. It is going through fragmentizers; some is getting into steel and some is being discarded in the fines and dirt that are thrown away.

One of the problems with fragmentizers is that they were designed essentially for processing cars and a scrapped car body, when it goes to the fragmentizer, has often had much of the non-ferrous metal removed from it. However, there are not enough scrapped cars around to feed the fragmentizers and they are taking quite a lot of white goods and other materials. In general, the non-ferrous content

is not removed, so a lot of this material is going through the existing scrap collecting system, and because of problems of sorting it we cannot recover it.

Professor ALEXANDER:
Those figures are insignificant compared to the 14 million tons.

Dr WHALLEY:
We cannot hope to get that 14 million tons. You are going back 30, 40 or 50 years. Much of it is deep in domestic refuse tips from which we cannot hope to recover it.

Dr A. F. BONFIGLIOLI
(University of Sussex):
I think it is quite important to introduce easier recycling as a criterion in product design and development. This really could lead to a more efficient recycling of old scrap but has to be considered from the very beginning of the product conception. Some progress has been made, the best-known example being beverage cans, and other developments are going on at present in this field. Another case has been that of some of the aluminium alloys in car bodies. But as far as I know recycling has not been generally introduced as a proper design criterion. Probably some incentive would be necessary. Could you comment on those points?

Dr WHALLEY:
I must say I was rather disappointed to see that two papers for this final session on the very subject of design for recycling had been withdrawn. Certainly this subject has emotive value. It is obviously desirable, to anyone interested in recycling, that the goods should be designed to make recycling easier, but some of the current trends towards the use of polymeric materials, composite materials, glass-reinforced plastic and so on are obviously going to make recycling in the future very much more difficult than it is now. I think it is a field that has not really been studied in any depth. It is widely considered to be a desirable aim, but I am not aware of very many people saying how we might go about doing it. I agree that it is something that we all ought to look at.

197

G. D. Bashford
Design for energy saving

Mr W. FAIRHURST
(Climax Molybdenum Co. Ltd):
One of the five features of the 1988 car to which you referred was the turbo-supercharged diesel. I believe that people looking at fuel economy and design of vehicles would note a very marked difference between the attitudes of the German and British designers towards the use of diesel engines for passenger cars. This affects metallurgy and many other things. Would you give your views as to why you do not mention the diesel in your presentation when there is such an apparent conflict of views with Continental designers on this?

Mr BASHFORD:
You are posing a leading question. I think I can readily say that British Leyland are working on diesel engines and the ultimate research work is pointing towards turbo-charged diesels. In other words, we are not idle in that particular scene.

Professor W. O. ALEXANDER
(University of Aston in Birmingham):
There are one or two points which underline what Mr Bashford said about this unreliability of information. I am not saying it affects his basic argument about aluminium versus steel but I have a suspicion that he has taken what I call exceptionally low values for aluminium for total energy required as compared with steel, or else he has got some very high values for steel. The other point is that I notice that the figure assumes a 60% saving in weight.

Mr BASHFORD:
No, 30% weight saving relative to a 60% gain in mpg terms. My example, which represented purely an example of what you could do on a body structure, did not of course relate all the other improvements in engine efficiency, aerodynamics and so on, or the reduction in

the total weight of the vehicle. I am saying that if you do all those things, it is possible to get something like a 60% gain in miles-per-gallon terms.

Professor ALEXANDER:
What I was going to elaborate was: when the car is fully laden (if it was in America, you would get 1·2 passengers per car) it seems to me that there are other operating factors which ought to be brought in right at the beginning of the design stage rather than operate a hypothetical car without any driver, etc.

Mr BASHFORD:
It is said that if you take your mother-in-law for a joyride it costs you 2 pence per gallon. The economy figures in my estimates are based on the ECE 15 test cycle which stipulates a test weight of vehicle, plus 120 kg, plus fuel tank half-full.

Mr P. J. KIDDLE
(Climax Molybdenum Co. Ltd):
On the question of materials and energy saving, I think we ought to be looking also at the source of that energy. We are talking all the time about energy in total, whereas I think our real problem is energy from oil and derived from oil. If we are talking in terms of nuclear power, then aluminium and the electricity available will be perfectly all right, whereas plastics could be in trouble because their energy content is derived from oil. I would like your opinion on this.

Mr BASHFORD:
I firmly believe that nuclear energy has certainly got to play a part in generating cheap electricity in order to overcome the certainty that oil will become scarce. This cheap electricity, I think will then be used for electrolysing water and forming hydrogen in order to propel our cars (I am now speaking about the next century). I agree that you have got to save oil; it is paramount.

Referring to Professor Alexander's remarks about materials energy, I want to pose a question of my own to this elite metallurgical audience. In order to calculate the amount of energy saved in building lighter cars, I have had to refer to figures published by various reliable sources which give the energy used in producing a given amount of aluminium or steel from mining of the ore to production of the finished sheet material. Unfortunately, the figures vary considerably, owing partly to the different premises and assumptions on which they are based. I find that this is particularly true of steel, for which I have found figures ranging from 9·5 to 18 kWh/kg in nine sources of information. For continuing research in this field of energy conservation it is very important that we have available average figures of unit energy per kg which it is felt reasonably represent the facts of current iron and steel production and of aluminium production. I would like to ask you as experts in metal production for your help in satisfying this need.

W. E. Duckworth
Changing patterns of materials use

Mr S. M. ST. JOHN
(Deloro Stellite):
Can you tell me whether, in your Table 5, the pounds quoted were constant pounds or adjusted for inflation? There appears to be a huge increase in the use of timber between 1966 and 1976.

Dr DUCKWORTH:
Those are pounds as they applied in 1966 and 1976. I have shared the problems of other authors in this conference of trying to get hold of really reliable statistics on a fairly uniform basis. If I can make a plea to The Metals Society Information Service, I think there is a marvellous opportunity here for the

Society to set up a really sound information base to collect and analyse statistics on the uses of materials. If we are going to keep track of what is happening we certainly need a far sounder base for comparisons than all the authors, I think, have found it possible to obtain.

Mr M. H. DAVIES
(Copper Development Association):
Dr Duckworth referred to the increasing cost of all forms of energy but of course he would readily agree, I am sure, that the cost of solar energy will not increase. Although the cost of conversion of that energy will change with time and with the use of materials, we shall see the increasing use of solar heat for water- and space-heating systems, low-grade process heat, and so on. That is not really the point that I want to make. I want to draw attention to an apparently innocuous remark in Dr Duckworth's preprinted paper: 'Under greatest threat are probably aluminium and copper because of price and political considerations. As Nutting has predicted, increasing price and scarcity will cause all metals except steel to settle into the role only they can fulfil'. I suggest that during the whole of this conference we have laid the bogey of scarcity as far as many of the metals are concerned, and particularly as far as copper is concerned. Of course, the patterns of use will change as relative prices alter. That is inescapable, but I would like to know what is meant by 'each metal settling into the role which only it can fulfil'. As far as copper and many other metals are concerned, the only role uniquely suitable to each individual metal is as a trace element in biological systems, about which I do not think we are particularly concerned here. We are concerned about the engineering use of materials and in that context we have to look at a situation in which the gloves are off. Competition is the name of the game. I would like to think that we will see increasing use of metals and all materials on the basis that I proposed yesterday and that other authors have also suggested: fitness for purpose and cost effectiveness. If that policy is followed, I am of course very confident about the future of copper.

Dr DUCKWORTH:
When I was talking about solar energy I agreed that, first of all, it certainly has a great

future for domestic heating, and for low-level energy sources, but I really do not see the construction of a solar power station generating the sort of electricity that we are going to need. The acreage of ground which would have to be covered is quite absurd.

I am glad you asked your question because I was not able to emphasize my point sufficiently in the presentation. What I am saying is that the major uses of the most common metals have by now been settled and they will remain in that rough pattern in the existing markets unless there are quite substantial price changes. When I started out to write this paper, I did try to begin by talking about significant changes in material uses which might take place in different design philosophies, but I very soon found that all the examples which I was able to obtain only affected the metals at the margin of their consumption; they do not significantly affect the overall relative pattern of materials in the hierarchy unless there is a sudden expansion in some particular market or an entirely new market opens up where all the materials are fighting on even ground. Let copper and aluminium fight it out in the electrical industry by all means; let steel and aluminium fight it out in furniture and windows by all means; but while the relative prices remain much what they are, there will not be significant changes that would show up on any of the graphs I have shown.

General discussion

THE CHAIRMAN: That takes us into the period of general discussion. I am going to use my position in the chair to make a few introductory comments which may help to stimulate this. It seems to me that there are three strands which could be profitably debated, although I do not want to confine comments to them.

The first is: you have heard from the proponents of the various metals a number of smug and complacent papers (I can say that because mine was not the least smug and complacent of the papers). You have heard the view put forward that resources are not really short and that the limitations of supply are more likely to be limitations in politics, finance, or perhaps especially in energy since everybody is agreed that energy is going to be short and there is going to be a massive increase in its price.

Secondly, there have been conflicting statements about substitution and replacement;

you have just heard Dr Duckworth's comments on that. The only example, we have had of massive replacement, I think, has been that of mercury given this morning. It would be interesting to have comments on that aspect; on why substitution and replacement appears to be so difficult; and, indeed, on whether it is desirable.

Thirdly, the point which has been made about the accuracy of statistics is extremely important for a learned society. A number of us have been quite rightly taken to task by Mr Archer for the rather loose way we have defined sources and reserves. Professor Alexander has raised the issue of old and new scrap and points have been made about the energy content of various materials. We stand perhaps a slight chance of getting policy right if we have the correct statistics. If these are inaccurate, as they are at the moment, clearly the question whether policy or strategy is right or not is going to be just a pure matter of chance.

Mr J. O. HITCHCOCK
(Engelhard Industries Ltd):

I ask myself the question: what have we achieved by this conference and what do we do about it? According to the speeches made yesterday and today, we do not have to worry too much about the availability of the raw materials, with the possible exceptions of mercury and zinc. If we look at what we are doing in terms of adequate metallurgical research, design, and recovery of secondary metals, we find ourselves in a fairly safe position, but I think the Society has to ask itself what course of action has resulted from this conference.

I have a suggestion: that if we let nature take its course, a lot of these things take care of themselves. The inevitable fluctuations in prices, which become adjusted to availability, take care of quite a lot of availability problems, so in effect I do not think we have too much cause for concern.

A DELEGATE:

I have made a particular study of the motorcar and would like to take up something Mr Bashford said. I think it is significant that most of the papers showing us how much energy is saved by replacing steel by aluminium have emanated from the aluminium companies.

The second thing I would like to say is that it is now firmly established that the factor defining lifetime of cars is not mechanical integrity but body corrosion. Were we to replace steel by aluminium (which can now be done, since the fabrication is no longer a problem) I submit that the lifetime of cars would be greatly increased. I also submit that the reason why manufacturers do not do this is that they know their replacement sales would be impaired. We cannot have it both ways.

Mr E. I. KAZANTSEV
(Economic Commission for Europe, Geneva):

As a representative of the ECE Geneva Branch and Industry Division, I should like to congratulate you on holding this conference on some very important topics, not only for the Societies but also for governments and the United Nations as well.

In the ECE we have more than 20 subsidiary bodies. One of them is the Iron and Steel Committee, in which, under the supervision of the representatives of the different governments, we have prepared three studies this year: 'The structural changes in the iron and steel industry', 'Increasing continuous processes in the iron and steel industries', and 'Steel scrap'. Those three studies will be published and on sale this year.

In my Division we have also begun two new studies, one on metallurgical coke, and the other on low-waste and non-waste technology, while the Energy Division will also publish a study this year, entitled 'Energy conservation in selected industries'. I have myself prepared the field for the iron and steel industry and shall be happy to admit Professor Alexander and anybody who is interested in these two new studies.

A DELEGATE:

May I firstly make a plea that contributions be collected and made available so that we have more time to study them, and can study them also in a variable framework.

My second point puts me in a position I do not relish and which is not familiar to me — that of devil's advocate — but one which I think someone must occupy. Let me begin by saying that there are many noble exceptions in this conference to the criticisms that I am going to imply; but having made that important caveat, let me assure you that I am not going to use the word 'smug' (both Chairmen used that, while Dr Duckworth selected the term 'arrogant'); I shall instead prefer the word 'conservative' which has a new respectability here and now. Collectively, the meeting seems to assume a relative stability for the future that, to put it mildly, may not be quite justified. When the title of 'Future Metal Strategy' is considered, I think it has been treated as though it had two sub-titles: firstly, that the theme is future metal strategy for the non-communist world overall — not for the EEC, not for the UK; and secondly, that it is future metal strategy assuming that the political and economic situation stays more or less on the present rails.

I would submit that in view of a number of factors, future metal strategy should be considered not by projection alone — that is the principal way in which it has been considered here — but also in terms of '*if*'. That, I feel, we have not done at all. Everyone is personally aware of the rapidity of political and economic change today; and everyone is or should be (perhaps it is not true of metal-lurgists) aware of the very long lead times connected with our industry. Certainly the miners are. Referring to my two sub-titles, may I add a couple of key words to those that have already appeared in the course of discussion? One of them, of course, is *South Africa*, which has been mentioned several times. I suppose it was not necessary to mention *South America* because, throughout the lifetime of all of us its history has been one of instability. Perhaps it would up-date my main point if I mentioned two names. One is Iran; the other I shall not name out-right but rather suggest that there might be in the not too distant future a new leadership and a new trend in the Labour Party.

Mr A. A. ARCHER
(Institute of Geological Sciences):
It has been fascinating for me as a geologist to attend this meeting and hear so many metallurgists speaking. The links are not, per-haps, as close as they should be between the geological profession and metallurgists. The general impression has been given (and con-firmed at the beginning of this discussion) that really there is no great scarcity, that generally we can afford to sit back and not worry.

I think that this conference has been con-fronted with two problems. One is the problem of definitions, to which we have already referred; we have to be very clear as to what we mean by reserves and resources. To that end, perhaps I should mention that there was a meeting called by the United Nations in January this year with a view to simplifying definitions and obtaining uniformity throughout the world. I believe this to be a step in the right direction. May I in that context issue a word of warning about always relying implicitly on the data published by the US Bureau of Mines? Because they are published in the English language and are readily available they are constantly quoted; but it is clear from our close and very friendly links with the Bureau that many of its figures are uncertain. They are not 'engraved on tablets of stone'!

The second problem which has emerged at this conference (and I say this with great diffidence) is that the contributors on each

of the metals have spoken from a particular point of view. I am reminded of what Mr Dowding, our Chairman, said yesterday: that it would have been entertaining if these con-tributors had spoken about the availability and use of substitute metals, rather than their own. We might then have received a different impression about scarcity!

I am sure that everyone here would agree, as would every geologist, that there is no question of absolute shortage of these materials in the world as a whole, with possibly some exceptions; mercury has been mentioned and one could think of other metals of which there may in fact be an absolute scarcity. But what we have to bear in mind (and here, not for the first time, I find myself in complete agreement with Professor Fleming) is that we must consider not only the matter of the quantities in the ground but also the political issues — not only those that have been highlighted during this conference but also those associated with mining investment. The mining houses are finding it increasingly difficult to decide where they can invest with the required degree of security of return. This is not pecu-liar to the developing countries of the world; it applies equally to many of the developed countries.

Finding ore bodies has become more diffi-cult. Lower grades are being worked and higher costs will apply in the future. So far, prices have in absolute terms stayed remark-ably constant for most metals during this century but we are, I am sure, moving into an era in which costs will increase — not just for energy reasons but for reasons of scarcity also.

This leads me to suggest a future strategy for metallurgists. First of all they will have to consider the relative supply difficulties for the various metals, taking into account all the factors that have been discussed at this con-ference — that is, their vulnerability in terms of supply. Then they will have to assess how important these metals are to industry; metal-lurgists are the only people who can carry out this study of criticality. Having put the metals in their order of relative vulnerability and criticality, they might examine what metallur-gical research should be undertaken to anticipate the problems that may arise in the future, perhaps at the beginning of the next century.

Finally, I very much regret that Sir Alan Cottrell is not present today since I would like to comment on two things that he said yesterday. Some of the delegates here know about the work we do in the Institute but there are others who do not. I am afraid this seems to include Sir Alan, who said that there is no domestic data base available for Government, whereas this is one of the things we are trying to provide. Secondly, because I believe the Department of Industry is not represented here, I cannot let it be thought that there is nobody in Government concerned with policy related to 'Future metal strategy'. On the contrary, there is very active work in progress.

Mr A. R. L. CHIVERS
(Lead/Zinc Development Association):
I am sorry that some of you should have got the impression that zinc reserves are inadequate. I mentioned yesterday a figure of 150 million tonnes of indicated reserves and the annual consumption in the world including the socialist countries is of the order of 6 million tonnes. I think there is quite enough zinc available for the preservation of motorcars by extending the use of galvanized steel. There is no need to develop aluminium technology in order to give car bodies better corrosion resistance. Zinc-coated steel has been widely used in North America for the past decade to counteract the vast quantities of salt and sand used to keep the roads clear in winter.

Dr H. ZIJLSTRA
(Philips Research Laboratories, Eindhoven)
I want to raise a point that has been mentioned once or twice in passing: that of the reclamation of metals from the sea. Virtually all metals occur in the sea, albeit in a very dilute form. There are potential techniques such as ion exchangers and membranes, but they are not considered feasible. What about biological systems? For example, iodine from seaweed is made in an economical way. We could think of making mercury from fish and it is known that there are iron deposits of microbial origin. Will it not be possible by genetic engineering to train bacteria to extract and concentrate certain elements from the ocean? Is there anyone who knows about efforts in that direction?

Mr G. D. BASHFORD
(British Leyland Cars):
I would like to react to the comment that was made about the aluminium companies being the main source of information regarding energy for steel. I have a list here which will give you some idea of that information. If you remember, I mentioned figures varying from something like 9·5 to 18 kWh/kg; the sources are *Metals and Materials*, by Mr Roberts in 1974, Singer and Dube, University of Swansea, in 1975, the Federation of Energy Administration and also associated with Singer and Dube, Mr Barnes in a presentation to the Institution of Mechanical Engineers, data from the *Engineering and Mining Journal*, and also the Iron and Steel Institute and International Iron and Steel Institute. There are also other figures from the United States Federal Energy Administration. So you can see that the figures are strongly steel-orientated, and not influenced by aluminium companies.

I agree with Mr Chivers's point about car preservation. The life cycle of cars, I believe, will have to be increased ultimately and that means, of course, that where steel is used in the outside environment it will have to be adequately protected. As you may remember, the VW slide that I showed, and the paper that was presented by VW, suggested adding something like 7 or 8 kg zinc to protect steelwork in that particular project.

THE CHAIRMAN:
I must now close the general discussion and call upon Professor Nutting; he was one of the fathers of this conference and I think it is appropriate that he should have the last word.

Closing address

Professor J. NUTTING
(revised on proof):

I should like to thank you for giving me the opportunity of making the closing remarks. The Working Party (Professor Alexander, Dr Hoare and myself) were very conscious of the problems likely to develop in organizing a meeting of this type. Optimism on the part of speakers talking about the metals on which their livelihood depends is perfectly natural. But in presenting their cases our speakers have had to argue their cause to an appreciative but not uncritical audience. As is almost usual, we have had insufficient time to cross-examine the plaintiffs, but a few weaknesses have been exposed.

It is perhaps reassuring that, except in the case of mercury, it seems very unlikely that real shortages of metallic resources are going to occur over the next twenty years. However, we must be prepared for changes in price relativities. Interestingly, we could be faced with a glut of manganese and possibly cobalt.

A real price increase in tin, copper, lead, and zinc would not be unexpected.

Writing now with hindsight, the political situation with respect to chromium (January 1980) looks more favourable than it did in May 1979. It is also a pity that we could not foresee in May 1979 that by January 1980 the price of gold would have almost doubled. But it all proves the wisdom of those who advise us never to make predications, particularly when their concern is for the future, and even more particularly the future price of metals.

The Metals Forum has now become an established reality and the data and views presented at the meeting and within this publication will provide a base for the development of our approaches to the next twenty years as members of an extensive metal-using society.

Author index

Page numbers marked with asterisk* are discussion contributions

205

Subject index

Page numbers marked with asterisk* are discussion contributions